INDUCED MUTATIONS
IN VEGETATIVELY
PROPAGATED PLANTS II

PANEL PROCEEDINGS SERIES

INDUCED MUTATIONS IN VEGETATIVELY PROPAGATED PLANTS II

PROCEEDINGS OF THE FINAL RESEARCH CO-ORDINATION MEETING
ON THE IMPROVEMENT OF VEGETATIVELY PROPAGATED CROPS AND
TREE CROPS THROUGH INDUCED MUTATIONS
ORGANIZED BY THE
JOINT FAO/IAEA DIVISION
OF ISOTOPE AND RADIATION APPLICATIONS OF ATOMIC ENERGY
FOR FOOD AND AGRICULTURAL DEVELOPMENT
AND HELD IN COIMBATORE, INDIA,
FROM 11 TO 15 FEBRUARY 1980

INTERNATIONAL ATOMIC ENERGY AGENCY
VIENNA, 1982

INDUCED MUTATIONS IN VEGETATIVELY PROPAGATED PLANTS II
IAEA, VIENNA, 1982
STI/PUB/519
ISBN 92–0–111182–7

© IAEA, 1982

Printed by the IAEA in Austria
April 1982

FOREWORD

A Co-ordinated Research Programme on the Improvement of Vegetatively Propagated Crops and Tree Crops Through Induced Mutations was sponsored by the IAEA between 1972 and 1980. It was based on the recommendations of an FAO/IAEA panel of experts who met in September 1972 and concluded that mutation induction was an appropriate tool for improving genetically crop plants that are not easily improved by cross breeding. The proceedings of that panel were published by the IAEA in 1973 under the title "Induced Mutations in Vegetatively Propagated Plants".

Research has been carried out to develop techniques for induction, selection and utilization of mutants in various vegetatively propagated crops. The programme has included a wide variety of plant species in tropical and temperate climates, such as sugarcane, potato, cassava, mulberry, citrus, bananas, apple, cherry, peach, grape, forage grasses, as well as a group of ornamental plants.

Researchers co-operating in this international programme met at two-year intervals to discuss their results and to plan the further line of work. Reports on the meetings held in Tokai (Japan) in 1974 and in Wageningen (Netherlands) in 1976 were available as documents IAEA-173 and IAEA-194 respectively.

The present book contains the reports presented by the project leaders at the final Research Co-ordination Meeting organized by the Joint FAO/IAEA Division of Isotope and Radiation Applications of Atomic Energy for Food and Agricultural Development, and held in Coimbatore, India, from 11–15 February, 1980. It also includes a review of the programme with conclusions and recommendations.

EDITORIAL NOTE

CONTENTS

SIGNIFICANCE OF IN VITRO ADVENTITIOUS BUD TECHNIQUES FOR MUTATION BREEDING OF VEGETATIVELY PROPAGATED CROPS*

C. BROERTJES
Foundation ITAL,
Wageningen,
The Netherlands

Abstract

SIGNIFICANCE OF IN VITRO ADVENTITIOUS BUD TECHNIQUES FOR MUTATION BREEDING OF VEGETATIVELY PROPAGATED CROPS.

It was investigated whether in vitro propagation techniques are of significance for the production of solid, non-chimeric mutants in mutation breeding programmes of vegetatively propagated crops. Irradiated explants of chrysanthemum, potato, begonia and carnation were used for the production of (adventitious) shoots and plantlets to determine the number and frequency of solid mutants and chimeras respectively. It was demonstrated that by the methods described high numbers of solid, non-chimeric mutants can be obtained and that the percentage of chimeras is comparable to the low figures reported after use of in vivo adventitious bud techniques. Consequently, the micro-propagation techniques seem very promising for the commercial plant breeder of vegetatively propagated crops.

INTRODUCTION

One of the bottle-necks in mutation breeding in vegetatively propagated plants is the chimera formation following the irradiation of multicellular apices present in buds of vegetative parts of the plant. It results in a number of unfavourable consequences, such as a reduced possibility to recognize mutated 'sectors', which prevents early selection and obliges the breeder to propagate all the material once or twice to arrive at complete periclinal chimeras (and a given percentage of solid mutants), before selection can be applied.

---

* This report is based upon research projects carried out in co-operative programmes between the Foundation ITAL (Broertjes: mutation breeding; Roest and Bokelmann: in vitro techniques), the Department of Plant Breeding (van Harten and Bouter: potato) and the Institute of Horticultural Plant Breeding (Sparnaaij and Demmink: carnation), all at Wageningen. The work on begonia and chrysanthemum was done at ITAL, in the case of begonia in collaboration with the Department of Horticulture at Wageningen (Doorenbos and Karper) and in the case of chrysanthemum with the interested support of a few commercial chrysanthemum breeders. The research was carried out in co-operation with the IAEA under Research Agreement No. 1485.

1

This situation can be improved considerably by using an in vivo adventitious bud technique which makes use of the phenomenon that the apex of adventitious buds often can be traced back to a single (epidermal) cell [1]. In the case of single-cell origin, adventitious plantlets will be either completely normal or completely mutated if the starting material was irradiated. The method allows early recognition and selection among the probably higher number of mutants than would other-wise have been obtained while their clonal propagation proceeds without diffi-culties. The potentialities of the method, also from a practical point of view, have been demonstrated in several ornamental crops such as *Achimenes* [2], *Begonia* [3] and *Streptocarpus* [4], of which commercial mutants were put on the market with-in a relatively short time after the first irradiation (approximately three years in the case of *Achimenes* and *Streptocarpus*). Full details can be found in Ref.[5].

Many plants can be propagated in vivo by adventitious plantlets [6], but many more cannot, and it thus seemed worthwhile to investigate the potentialities for mutation breeding of in vitro propagation techniques, in the sense that solid, non-chimeric plants can be produced by using irradiated explants. We, therefore, have selected a number of crops known to mutate easily because of their heterozygosity and in which the development of an in vitro propagation method through adventitious shoots was expected to be relatively easy. The crops selected were chrysanthemum, potato, begonia and carnation. The production of commercially interesting mutants was not an objective of this programme.

CHRYSANTHEMUM

The *Chrysanthemum morifolium* cultivars Bravo and Super Yellow were used to develop in vitro propagation techniques and to study the effect of explant type, explant size, medium-components, etc., upon speed of adventitious shoot formation and number of shoots. The main conclusions are as follows (full details in Ref.[7]):

Pedicel-segments very rapidly (10—14 days) produce adventitious shoots, mainly directly from epidermal cells. Other explants (petals, explants of the capitulum or of the leaf) also produce shoots but fewer, later (3—4 weeks) and usually via callus.

The effect of various factors (explant size, various cytokinins and auxins, sucrose, vitamins, micro- and macro-elements) upon speed of shoot formation and number of shoots is very dependent on the cultivar used. Under optimum conditions, pedical explants of Bravo developed considerably more adventitious shoots than those of Super Yellow. This was confirmed in a later experiment with about 10 different cultivars where the number of shoots varied from 15—50 per shoot-forming explants.

The best rooting of excised shoots was obtained by incubating them under sterile conditions in the same medium as used for shoot formation but supplemented with 2% sucrose and 10^{-7} g/ml IAA [7].

The method described above was used to investigate its significance for mutation breeding. Various explant-types of the cultivar Bravo were irradiated with different doses of X-rays before use in vitro. The shoots formed were rooted and grown to maturity. Flower colour mutants were mainly used to determine, by visual inspection, whether solid-looking mutant plants or chimeras were obtained. The results indicate that:

All mutants were solid-looking (except one). A few of the mutants were propagated by cuttings and re-irradiated with a high dose of X-rays. The fact that no cell-tissue rearrangement and uncovering was observed is strong evidence that the mutants were not only solid-looking, but were non-chimeric genetically homogeneous mutant genotypes.

The optimum dose lies around 800 rad, considering the number of shoots per shoot-forming explant and mutation frequency[1].

Identical mutant genotypes were found rather frequently, always derived from one explant. In one case even all 22 shoots of one explant were identical mutant genotypes. This indicates that a mutated cell grew out into a multi-apical meristem on which several genetically identical shoots were produced.

The final conclusion is that the use of irradiated explants, notably those of young pedicels, for the production of adventitious shoots in vitro, is a valuable tool to obtain non-chimeric and genetically homogeneous mutant genotypes within a comparatively short time period. The advantages and disadvantages of this method for mutation breeding of chrysanthemum are discussed in Ref.[9]. Some of the advantages are:
 Fast and comparatively easy method
 The mutants are stable
 Easy and early screening
 Easy vegetative propagation
 Mutants can be used in cross-breeding
 Higher mutation frequency?
 Wider mutation spectrum?

POTATO *(Solanum tuberosum* L.)

To investigate whether similar results could be obtained in potato, a co-operative project was started in which the Foundation ITAL developed in vitro

---

[1] 1 rad = 1.00×10^{-2} Gy.

TABLE I. NUMBER OF ADVENTITIOUS PLANTLETS PRODUCED IN VITRO
FROM DIFFERENT EXPLANT TYPES OF POTATO, *Solanum tuberosum* L.,
FOLLOWING VARIOUS X-RAY DOSES

Object number	Explant type	X-ray dose (rad)[a]	No. of explants used	No. of explants producing adventitious shootlets	No. of adventitious plantlets obtained
1	Rachis and petiole	0	20	18	158
2	Rachis and petiole	1500	100	87	834
3	Rachis and petiole	1750	120	81	538
4	Rachis and petiole	2000	120	79	461
5	Leaflet blade	0	20	18	203
6	Leaflet blade	2250	100	61	421
7	Leaflet blade	2500	120	66	404
8	Leaflet blade	2750	120	47	148

[a] 1 rad = 1.00×10^{-2} Gy.

TABLE II. FREQUENCY OF SOLID 'MUTANTS' AND OF CHIMERAS OF
POTATO, *Solanum tuberosum* L., IN MV_1 PRODUCED IN VITRO, FOLLOWING
IRRADIATION OF DIFFERENT EXPLANT TYPES

Object number[a]	Aerial part of plants			Subterranean part of plants		
	No. observed	No. of 'mutants'[b]	No. of chimeras	No. observed	No. of 'mutants'[b]	No. of chimeras
1	157	35 (22%)	0	153	52 (34%)	1 (0.7%)
2	772	170 (21%)	8 (1%)	756	343 (45%)	7 (1%)
3	504	269 (53%)	2 (0.4%)	487	214 (44%)	13 (3%)
4	442	266 (62%)	4 (1%)	431	244 (57%)	10 (2%)
5	202	18 (9%)	1 (0.5%)	201	28 (14%)	7 (3%)
6	394	237 (62%)	0	381	180 (47%)	11 (3%)
7	377	227 (66%)	1 (0.3%)	344	207 (60%)	1 (0.3%)
8	134	74 (62%)	0	120	70 (58%)	1 (0.8%)

[a] See Table I.
[b] These columns include all slightly deviating plants besides the obvious mutant ones. In the MV_2 generation a number of the first category were normal, non-mutated plants.

adventitious bud techniques [10, 11] and determined the radiosensitivity of the material, whereas the Department of Plant Breeding (Agricultural University, Wageningen) tested the potentialities of the method for mutation breeding [12].

To develop in vitro adventitious bud techniques, just expanded compound leaves of the cultivar Désirée were used to prepare explants of the rachis and petioles, and also the upper pair of lateral leaflets were detached and used as the explant source.

After sterilization 0.5 cm long rachis and petiole segments were excised and divided longitudinally into two similar explants before placing them horizontally with the wounded side on a culture medium. From the leaflets 0.5 cm square leaflet-blade explants were taken from the midrib vein area and placed with the lower side in the test tube with the same culture medium. We used a solidified Murashige-Skoog medium, complemented with sucrose, benzylamino purine, indoleacetic acid and gibberellic acid. The tubes with the explants were placed in a growth cabinet at a temperature of 20°C, a relative humidity of 70–80% and a light intensity which varied from 1000–10 000 lx, from the bottom shelf to the top shelf, respectively. The day length used was 14 h. After shoot development, shootlets of at least 1 cm in length were excised and subcultured for rooting on a somewhat different medium. After rooting, the plantlets thus obtained were transplanted to soil and established.

Similarly with unirradiated material, irradiated leaves were used for explant production. Different X-ray doses were applied, such as 1.5, 1.75 and 2 krad for the rachis and petiole explants, and 2.25, 2.5 and 2.75 krad for the leaflet-blade explants. These doses are around the optimum one, determined earlier by applying a much longer series of X-ray doses.

Over 3000 M_1 plantlets, from irradiated and partly also from unirradiated material, were transferred to our colleagues at the Department of Plant Breeding for further growth and study (Table I).

The significance for mutation breeding

The MV_1 plants were transferred in 13 consecutive series to the Department of Plant Breeding and observed during several months in 1977. These various delivery times made the comparison between the first developed plantlets and the last ones very difficult. Apart from clearly mutated plants, most of the others, notably those from explants exposed to high X-ray doses, had a slightly abnormal appearance. This is partly due to radiation-induced non-genetic damage but it also seems to result from the effect of greenhouse conditions and from the propagation method used, in view of the rather high number of such plantlets ('mutants') in both controls. As was expected it is not, or to a lesser degree, found in the MV_2 and MV_3 [5, p.60]. Lower numbers and percentages resulted as indicated in Table II, columns 3 and 6, thus it could not be concluded from these data to what degree the in vitro method as such was 'mutagenic'.

From the MV_1 data the following conclusions could be drawn:

1.   The number and percentage of chimeras is low, on average 1.3%, if mutations in the aerial and subterranean parts of the plants are combined as a parameter (habitus, size, shape, colour or hairiness of plants and leaves; shape and skin colour of the tubers). Whether the non-chimeric mutants are periclinal chimeras or solid mutants is hard to say without further investigation. Most of the yellow-skinned tuber mutants, however, indicate that the mutants were solid (48 out of 51). (Periclinal chimeras would be red-yellow variegated; solid mutants are completely yellow, as was the case in most of our mutants, whereas unmutated plants have completely red tubers.)

2.   The number of frequency of the mutants is high, also if it is taken into account that a number of aberrant MV_1 plants turned out to be normal in the MV_2 and MV_3. Irradiated rachis and petiole explants produced ultimately 68.2% mutants. (The control, however, contained as many as 50.3%.) The figures for irradiated leaflet-blade explants were 85.9% and 12.3% respectively. The higher mutation frequency in the control of the first type of explant cannot be explained adequately.

The MV_2 were planted in 1978 and the MV_3 in 1978. They largely confirmed the preliminary conclusions and the final conclusion therefore is that large-scale in vitro production of adventitious plantlets of potato, *Solanum tuberosum* L., is available and can be used to produce large numbers of mainly solid, non-chimeric mutants, if leaves are irradiated prior to the preparation of explants from the rachis, petiole or the leaflet blade. In view of the high percentage of aberrant MV_1 and MV_2 plants it does not seem attractive to use the in vitro methods described as a true-to-type vegetative multiplication method. More detailed information can be found in Ref.[12].

BEGONIA

To investigate the possibilities in begonia use was made of the clone 'SO 1', kindly offered to us by the Department of Horticulture (Agricultural University, Wageningen). This clone is a cross-breeding product between the tetraploid yellow-flowering tuberous begonia cultivar *Bertinii compacta* Sonnenschein and the diploid red flowering *Begonia socotrana.*

This triploid and sterile clone 'SO 1' was successfully used for the production of thousands of adventitious shoots (using an in vivo propagation method with detached leaves) of which the majority (99%) were solid, non-chimeric mutants.

One of these mutants, a yellow-flowering genotype, has been introduced into commerce under the name of 'Tiara' [3].

At ITAL, we used the clone SO 1 to develop an in vitro adventitious bud technique and to investigate whether solid mutants could be obtained [13, 14]. In addition, the cultivar Schwabenland Red was used, because it easily produces spontaneous mutants of which several have been commercialized.

The method of in vitro propagation is, in short, as follows (full details are to be published in English): leaf blade explants of about 0.5 cm^2 are taken from leaves that are just mature, decontaminated with 50% calcium hypochlorite and placed horizontally on the medium under sterile conditions. The best medium contains agar 8 g/l, sucrose 30 g/l to which the organic and inorganic components of the Murashige and Skoog medium are added, supplemented with benzyladenine and naphthalene acetic acid, 1 mg/l. The pH of the medium is adjusted to 5.8 with a 1N KOH-solution.

Adventitious bud formation takes place after about one month under normal culture conditions. For reasons to be discussed later, leaf explants of these adventitious shoots are used for further shoot production using the method described above. As soon as sufficient shoots are produced, parts of the explants with one or a few adventitious shoots are transplanted to sterilized soil for further growth and rooting, which takes place within two to three weeks. This is a very critical phase of the whole procedure and it is important to take every possible measure to avoid damping-off of the fragile shoots and plantlets. As soon as rooted the plantlets are grown to maturity.

To investigate the significance of this in vitro propagation method for mutation breeding, plants were grown from irradiated explants. The doses ranged from 0.5 to 3 krad X-rays. The optimum dose lies around 1.5—2 krad, taking speed of shoot formation, number of adventitious shoots per explant and mutation frequency as parameters.

It is difficult to decide upon the degree of chimerism, if any, in the first set of adventitious shoots. Because these shootlets are very compact, small and fragile it is impossible or at least impracticable to try to separate and root individual adventitious shoots and we, therefore, introduced a second round of taking explants and producing adventitious shoots as described above. By doing this and by restricting the number of explants to one per original adventitious shoot (as far as possible) a rather large number of mutants were produced, all being solid and non-chimeric, apart from one (full details are to be published).

The final conclusion is that the micro-propagation technique described above is one way to produce large numbers of non-chimeric mutants besides the in vivo adventitious bud technique, using detached leaves, as used and described by Doorenbos and Karper [3].

CARNATION

Carnation, *Dianthus caryophyllus* L., was the last crop we used to investigate the usefulness of an in vitro propagation technique by which (adventitious) shoots and plantlets are produced.

The only possible micro-propagation method was to use nodal stem explants which produce rather readily a great number of axillary shoots and, in addition, most probably also adventitious shoots. Explants of six different cultivars were irradiated with different doses of X-rays in 1977, and over 2200 plantlets obtained from this irradiated material were transferred in 1978 to the Institute of Horticultural Plant Breeding for further cultivation and evaluation for the mutation breedings aspects in 1979 and following years [15].

To our surprise most flower colour and other mutants were solid-looking, but further vegetative propagation and reirradiation of some of the mutants is necessary to arrive at a final conclusion on the nature of the mutants.

Our preliminary conclusion, however, is that this method seems to be a rather fast and easy way to produce large numbers of (complete or periclinal?) mutants which is already an attractive proposition, especially if one is interested in early selection, e.g. for non-directly visible characteristics such as disease resistance (full details are to be published).

DISCUSSION AND CONCLUSION

From the data obtained on chrysanthemum, potato, begonia and probably also on carnation, it can be concluded that any micro-propagation technique by which adventitious shoots are produced is a valuable method to obtain solid, non-chimeric mutants in crops where such a method is available or can be developed. Since the number of crops in which micro-propagation is reported and described in the literature has increased dramatically during the last decade, it seems that mutation breeding in most crops is no longer restricted by the limiting consequences of chimera formation.

This seems, at first sight, to be contradicted by the successful mutation breeding results obtained in several crops in which rooted cuttings or other plant parts with existing buds are used for irradiation and the subsequent production of sports, mostly periclinal chimeras. Even in a crop such as chrysanthemum where mutation breeding has been very successful in producing flower colour and other mutants [16], there is a tendency to favour the use of the in vitro adventitious bud technique in the production of non-chimeric mutants above the usual method of irradiating (rooted) cuttings. This relates to the problem of stability of periclinal chimeras or sports during vegetative propagation, and the fact that periclinal chimeras generally cannot be used as cross-breeding parents, in contrast to solid mutants.

Moreover, since true-to-type micro-propagation is being used commercially for a rapidly increasing number of crops, it is evident that genetically homogeneous seedling cultivars or solid mutants are less prone to stability problems compared with the micro-propagation of chimeras. This undoubtedly is the case if adventitious buds are involved, but also in the case of a meristem culture or the use of very tiny shoots for 'in vitro' propagation, the formation of homogeneous (adventitious) shoots cannot be excluded, as well as the 'normal' percentage of instability.

For the production and selection of mutants with non-directly visible characteristics (yield, disease-resistance, decreased temperature need, etc.), one could imagine that solid mutants are to be preferred. In view of the impossibility of recognizing in time cases of uncovering or other cell-tissue rearrangements during commercial vegetative propagation and the subsequent unacceptable spoilage of the genetic homogeneity of the clone, it is necessary to have solid (mutant) geno-types to maintain the genetic integrity of the clone during propagation.

Solid, non-chimeric (mutant) genotypes will also be produced by the use of certain in vitro techniques such as protoplast or single-cell culture, and more notably cell suspension culture, and the subsequent regeneration of callus and adventitious shoots and plantlets. Such techniques are within reach for the production of biochemical and other mutants, e.g. resistance to toxins of certain plant diseases. It will enable us to handle millions of cells and to screen for rare (dominant) mutational events such as is already done in microorganisms.

A few words are added on (1) the use of DTT (dithiothreitol) to improve the mutation spectrum of vegetatively propagated crops; and (2) the comparison of diploids and their respective autotetraploids of vegetatively propagated crops for mutation breeding.

The first subject was discontinued because of disappointing results [17]. The second subject is still being researched but so far without new data or aspects as was reported on earlier [18].

REFERENCES

[1] BROERTJES, C., KEEN, A., Adventitious shoots: do they develop from one cell? , Euphytica 29 1 (1980) 73.
[2] BROERTJES, C., Mutation breeding of *Achimenes,* Euphytica 21 (1972) 48.
[3] DOORENBOS, J., KARPER, J.J., X-ray induced mutations in *Begonia* X *hiemalis,* Euphytica 24 (1975) 13.
[4] BROERTJES, C., Mutation breeding of *Streptocarpus,* Euphytica 18 (1969) 333.
[5] BROERTJES, C., van HARTEN, A.M., Application of mutation breeding methods in the improvement of vegetatively propagated crops: An interpretive literature review, Elsevier, Amsterdam (1978) 316.
[6] BROERTJES, C., HACCIUS, B., WEIDLICH, S., Adventitious bud formation on isolated leaves and its significance for mutation breeding, Euphytica 17 (1968) 321.

[7] ROEST, S., BOKELMANN, G.S., Vegetative propagation of *Chrysanthemum morifolium* Ram. in vitro, Scientia Hort. **3** (1975) 317.

[8] BROERTJES, C., ROEST, S., BOKELMANN, G.S., Mutation breeding of *Chrysanthemum morifolium* Ram., using in vivo and in vitro adventitious bud techniques, Euphytica **25** 1 (1976) 11.

[9] BROERTJES, C., "The improvement of *Chrysanthemum morifolium* Ram. by induced mutations", Proc. Eucarpia meeting on Chrysanthemums, Littlehampton, England, October 1978 (1979) 93.

[10] ROEST, S., BOKELMANN, G.S., Vegetative propagation of *Solanum tuberosum* L. in vitro, Potato Res., **19** (1976) 173.

[11] ROEST, S., BOKELMANN, G.S., In vitro adventitious bud techniques for vegetative propagation and mutation breeding of potato (*Solanum tuberosum* L.). I. Vegetative propagation in vitro through adventitious shoot formation, Potato Res. **23** (1980).

[12] VAN HARTEN, A.M., BOUTER, H., BROERTJES, C., In vitro adventitious bud techniques for vegetative propagation and mutation breeding of potato *(Solanum tuberosum* L.). II. Significance for mutation breeding (in preparation).

[13] ROEST, S., BOKELMANN, G.S., Vermeerdering van begonia in kweekbuizen. Vakbl. Bloem. **35** (1980) 116.

[14] ROEST, S., BOKELMANN, G.S., Mutatieveredeling van begonia via een in vitro adventief spruitmethode, Vakbl. Bloem. **35** 6 (1980).

[15] ROEST, S., BOKELMANN, G.S., Vermeerdering anjer in kweekbuizen, Vakbl. Bloem., **34** (1979) 38.

[16] BROERTJES, C., KOENE, P., VAN VEEN, J.W.H., A mutant of a mutant of a mutant ..... The irradiation of progressive radiation-induced mutants in a mutation breeding programme with *Chrysanthemum morifolium* Ram., Euphytica **29** 2 (1980).

[17] BROERTJES, C., "Is DTT a means to improve the mutation spectrum of vegetatively propagated plants? ˮ Improvement of Vegetatively Propagated Plants and Tree Crops through Induced Mutations (Proc. Second Research Co-ordination Meeting, Wageningen, 17–21 May 1976), IAEA-194 (1976) 13.

[18] BROERTJES, C., "Mutation breeding of autotetraploid *Achimenes* cultivars", Improvement of Vegetatively Propagated Plants and Tree Crops through Induced Mutations (Proc. Second Research Co-ordination Meeting, Wageningen, 17–20 May 1976), IAEA-194 (1976) 1.

PROGRESS IN MUTATION BREEDING OF APPLES (*Malus pumilla* Mill.) AT LONG ASHTON RESEARCH STATION, BRISTOL, UNITED KINGDOM*

C.N.D. LACEY, A.I. CAMPBELL
Long Ashton Research Station,
Bristol, Avon,
United Kingdom

Abstract

PROGRESS IN MUTATION BREEDING OF APPLES (*Malus pumilla* Mill.) AT LONG ASHTON RESEARCH STATION, BRISTOL, UNITED KINGDOM.

A brief summary of the techniques used to produce mutants from over 4000 scions of fifteen cultivars of apple is given. Basic work within that programme has given information indicating the best dose for mutant production for the various cultivars (approximately the LD_{50} for the growth of the MV_1). It has been shown that selection for greater damage at the MV_1 generation can lead to a higher proportion of mutants in later generations, but also that the more severe mutations are less likely to be of commercial interest. Half of the selected mutants resulting from this programme appear to be fairly stable so far and are performing well in large-scale trials. Mention is made of techniques which resolve the chimeral make-up of the remaining potentially useful mutants.

INTRODUCTION

Apple trees are, in theory, ideal subjects for a mutation breeding programme. They are highly heterozygous so that induced mutants may show phenotypically without the necessity for a seed propagated generation. They are normally vegetatively propagated, so that a new mutant can be easily multiplied and, in some countries, statutory schemes exist for the dissemination of new clones to the industry. Examples are the Dutch N.A.K.B. scheme and the English EMLA scheme, both originally designed for the distribution of virus-free and pomologically superior clones.

Propagation of these healthy clones leads to extremely uniform rows of plants in the nursery, making the selection of mutants easy, particularly for those with changes in vegetative characters. The apple industry in the United Kingdom and other fruit growing areas of the world is very conservative which, while making the acceptance of an entirely new cultivar more difficult, eases the introduction of

---

* Research carried out in co-operation with the IAEA under Research Agreement No. 2033.

11

TABLE I. SOME DEFECTS IN POPULAR APPLE CULTIVARS GROWN IN THE UNITED KINGDOM WHICH COULD BE ALLEVIATED BY MUTATION BREEDING

Cultivar	Defect
Cox's Orange Pippin	Low fertility. Poor skin colour. Short harvest period. Small fruit. Susceptible to disease.
Bramley's Seedling	Too vigorous. Uneven shaped fruit.
Crispin (Mutsu)	Too vigorous. Fruit too large.
Holstein	Too vigorous. Fruit too large.
Tydemans Worcester	Too much bare wood.
Discovery	Slow to start cropping.
Golden Delicious	Russet under U.K. conditions.
Kidd's Orange Red	Poor skin finish.
Spartan	Poor skin colour.

new mutants of traditional cultivars. Despite their popularity with the consumer all the commonly grown apple cultivars have defects from the growers' point of view, many of which should be amenable to mutagenesis (Table I).

MATERIALS AND METHODS

Dormant graftwood of a selected, normally EMLA status, virus-free clone of the scion cultivar to be treated is collected in the winter and cut into lengths to fit the radiation source to be used and suitable for grafting. This material is termed the MV_0 generation, and is subjected to a measure dose of radiation under controlled conditions, normally with the propagation material immersed in water (see below). The treated material is then grafted on to pot-grown, virus-free clonal rootstocks (normally MM106 as it produces most uniform plants) and grown in a cold glasshouse. This ensures the greatest possible survival of the plants, and shoots are produced from at least the distal bud of each surviving graft. These shoots are the MV_1 generation, and show damage or growth reduction dependent largely on the amount and type of radiation used.

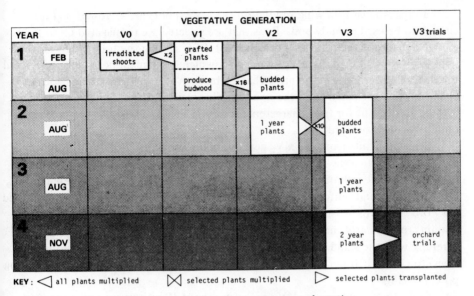

FIG.1.   Mutation breeding programme schematic.

TABLE II.   NUMBER OF SCIONS OF APPLE CULTIVARS TREATED IN THE LONG ASHTON MUTATION BREEDING PROGRAMME

Cultivar	No. of scions irradiated
Cox	862
Bramley	444
Queen Cox	420
Crispin (Mutsu)	593
Spartan	365
Malling Kent	364
Suntan	391
Golden Delicious	144
Gala	140
+ Six other cultivars < 100 each	480
Total	4203

By the August following irradiation the MV_1 shoots are large and mature enough to be used as budwood. Buds that are selected are propagated on to field-grown clonal rootstocks (again normally MM106). During the following year these buds grow on to form the MV_2 generation, which is the main population subjected to selection. Mutants of vegetative characters, such as dwarf growth and foliage changes, can be selected from the one-year-old (maiden) trees. Where selection is to be carried out for fruit characters the young trees must be transplanted to a semi-permanent orchard site.

Selected plants from the MV_2 generation are propagated again by summer budding to produce the MV_3 generation. This provides a check on the stability and reality of the mutations, helps identify mericlinal chimeras and, where the MV_3 trees are stable and uniform, provides material for replicated trials. These replicated MV_3 orchard trials are medium to long term, i.e. five to ten years, as promising mutants, particularly those selected for vegetative characters, have to be compared carefully with the standard tree of the same cultivar (produced alongside the MV_3 trees) to check their performance for all important characters, particularly cropping potential (Fig.1).

Clones selected from the MV_3 orchards (about 1% of the mutants found) are then re-propagated on a large scale (hundreds of plants) to produce an MV_4 population of each clone. These are measured to check on uniformity, and good uniform clones go forward to final commercial trials, on Ministry of Agriculture Experimental Horticulture Stations, at the National Fruit Trials and on commercial growers' farms.

As with all top fruit breeding programmes it is impossible to wait for the results of one experiment before starting the next, so for the past 11 years some irradiation treatments have been carried out every year, and every stage of the programme is in progress at the same time.

Eighty-three separate irradiation experiments using 15 different cultivars of apples as well as apple rootstocks, pear, plum, cherry and soft fruit cultivars and ornamental shrubs and trees have been undertaken. The initial treatments have been given to populations ranging from tens to hundreds of scions or cuttings. The main apple cultivars irradiated and the number of scions involved are listed in Table II.

In most cases practically all the scions treated survived and the aim with our programme is to apply enough radiation to reduce growth to about 50% of the control in the MV_1 generation (see below). Thus when the MV_1 shoots are used to produce the MV_2 generation, very large populations can result, for instance some 20 000 MV_2 Cox and Queen Cox trees have been examined, and 6000 Bramley's Seedling. Altogether there are still over 15 000 MV_2 trees planted and available for selection, mostly in commercial growers' orchards. As the majority of the plants are not seriously affected by the radiation treatment, growers have, in general, been pleased to take the one-year-old trees. These are the MV_2 genera-

WILLOW

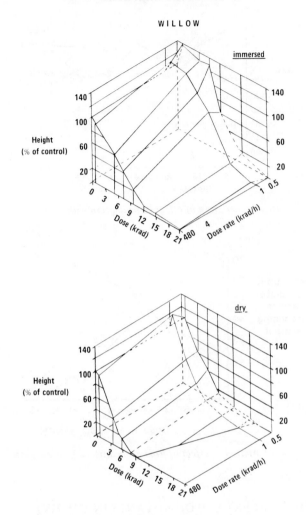

FIG.2. *Effect of irradiation conditions on the primary effect of radiation.*

tion raised in the Long Ashton nurseries and with the obvious vegetative mutants removed for our own trials. The trees remain the property of the research station, but all fruit in excess of experimental needs (normally only a few samples are needed) is the property of the grower.

The trials of MV_3 clones on the station (so far only selected from our own nurseries) at present include some 3500 apple trees of about 500 mutant clones. This does not represent the entire yield of mutants from the trials listed in Table II, as many MV_2 populations have yet to yield their full quota of mutants, and other mutants have already been discarded.

(text cont. on p.21)

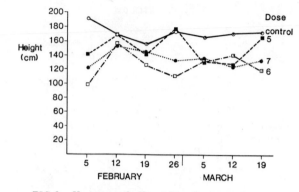

FIG.3.   Variation of effect of irradiation with time.

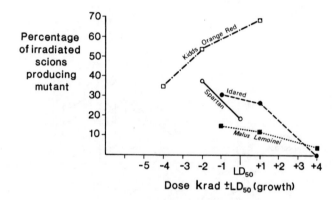

FIG.4.   Variation of effect of irradiation with cultivar and dose.

TABLE III.   THE PERCENTAGE OF MUTANTS IN THE MV_2
GENERATION AFTER THE IRRADIATION OF FOUR
CULTIVARS OF APPLE

	Approximate dose (krad)	0	5	7	10	Overall percentage
Cultivar	*Malus lemoinei*	0	16	13	33	16
	Spartan	0	37	63	a	43
	Idared	0	33	29	0	31
	Kidd's Orange Red	0	41	54	72	56

[a] All treated scions killed.

TABLE IV. PROPORTION OF DIFFERENT CATEGORIES
OF THE MV_1 BUDSTICKS AND THEIR SIZES

MV_1 shoot description	Long 1	Long double 2	Short 3	Short double 4	Short treble 5	Dwarf 6	Total or average*
No. (MV_1)	93	46	52	21	12	20	244
Percentage	38	19	21	9	5	8	100
No. buds/ stick	20^c	26^a	14^d	20^{bc}	31^a	10^d	20*
No. buds (MV_2)	1824	1212	736	414	368	204	4758

Figures in the row (*) followed by the same letter are not significantly different at the 5% level.

TABLE V. EFFECT OF DIFFERENT TYPES OF MV_1 BUDSTICK ON
THE NUMBERS OF MUTANTS PRODUCED IN THE MV_2

MV_1 shoot description	Long 1	Long double 2	Short 3	Short double 4	Short treble 5	Dwarf 6
No. trees (MV_1)	1824	1212	736	414	368	204
No. mutated trees	146	94	125	74	71	51
Percentage mutated trees	8.0^b	7.8^b	17.0^a	17.9^a	19.3^a	25.0^{a*}
No. mutant types	31	18	15	9	9	9
Percentage mutant types	1.7	1.5	2.2	2.6	2.6	4.4

Figures in the row (*) followed by the same letter are not significantly different at the 5% level.

TABLE VI. THE TYPE, POSITION AND NUMBER OF MUTATED BUDS ON MV$_1$ SHOOTS OF BRAMLEY'S SEEDLING APPLE

Description of MV$_1$ shoot	Long single	Long double	Short single	Short double	Multiple	Dwarf	Total
Code	1	2	3	4	5	6	
	No.(%)	No.(%)	No.(%)	No.(%)	No.(%)	No.(%)	
Runs of mutated buds							
Lower section	2(11)	2(16)	0 0	0 0	0 0	0 0	4
Middle section	3(16)	2(15)	2(20)	0 0	0 0	0 0	7
Top section	3(16)	0 0	0 0	0 0	0 0	1(20)	4
One complete shoot or more	3(16)	3(23)	6(60)	4(67)	2(40)	3(60)	21
Sub-total	11(58)	7(54)	8(80)	4(67)	2(40)	4(80)	36
Single mutated buds							
Buds 1–5	6(32)	4(31)	1(10)	2(33)	3(60)	1(20)	17
Buds 6–top	2(11)	2(15)	1(10)	0 0	0 0	0 0	5
Sub-total	8(42)	6(46)	2(20)	2(33)	3(60)	1(20)	22
Total	19	13	10	6	5	5	58

TABLE VII. CLASSIFICATION OF 236 VEGETATIVE MUTANTS DERIVED FROM 747 SCIONS OF GAMMA-IRRADIATED COX'S ORANGE PIPPIN

Class	Dwarf			Weak		Tall	Weeping	Leaf			Bark
	Slight	Medium	Extreme	Medium	Extreme			Twisted	Thick	Early fall	
Number found	86	72	31	20	2	2	10	5	1	4	3
Percentage of all mutants	36.4	30.5	13.1	8.5	0.8	0.8	4.2	2.1	0.4	1.7	1.3

TABLE VIII. REASONS FOR REJECTING APPARANT MUTANTS AT THE MV$_3$ STAGE

Mutant type	Original number of clones	Total % rejected	Reason for rejection (%)[a]				
			Similarity to control	Probable chimera	Sterile (no flowers)	Poor fruit quality	Other reasons
Slight dwarf	86	64.0	28.6[a]	12.8[b]	21.0[ab]	31.9[a]	5.7[c]
Medium dwarf	72	56.9	26.6[a]	13.8[b]	21.8[ab]	33.0[a]	4.8[c]
Extreme dwarf	31	67.8	21.1[b]	26.3[b]	36.8[a]	10.5[b]	5.3[b]
Weak	22	72.7	5.9[b]	11.8[b]	64.7[a]	17.6[b]	0.0[b]
Weeping	10	80.0	12.5[ab]	25.0[ab]	12.5[ab]	50.0[a]	0.0[b]
Others	15	87.7	24.3[a]	7.1[a]	42.9[a]	28.6[a]	7.1[a]
Overall rejected	154	100	11.4	14.7	29.5	28.9	4.5
Percentage of all mutants		65.3	14.6	9.6	19.2	18.9	2.9

[a] Numbers in any row followed by the same letter are not significantly different at the 5% level.

TABLE IX. DERIVATION OF MUTANTS SELECTED FOR
ORCHARD TRIALS

Mutant type	Number of mutants	Percentage of original mutants of that type
Slight dwarf	31	36.0
Medium dwarf	31	43.1
Extreme dwarf	10	32.3
Weak	6	27.3
Weeping	2	20.0
Others	2	13.0
Total	82	34.7

As stated above, the best clones from these trials are further multiplied to produce trees for commercial trials (MV$_4$). So far this stage has only been reached for those mutants of Cox's Orange Pippin and Bramley's Seedling originally irradiated in 1969–71.

RESULTS AND CONCLUSIONS

Many results from this programme have been published elsewhere (see Bibliography), so the findings will only be summarized here.

Initial effects of radiation

The initial effects are variable in terms of the amount of primary effect caused by any particular dose. The main factor that causes this variation is the rate of application of the gamma radiation, even when comparing acute treatments. The conditions under which the radiation is applied also causes variations in response. As explained elsewhere [1], plant material is almost always irradiated while it is submerged under water to avoid the 'build-up effect', but this in itself makes the radiation less effective, so that a larger dose has to be given. This water protection effect seems independent of the oxygen effect. Figure 2 shows two graphs showing this effect for willows (*Salix* spp). Apple cultivars also vary in their sensitivity to irradiation, even within the diploid and triploid groups.

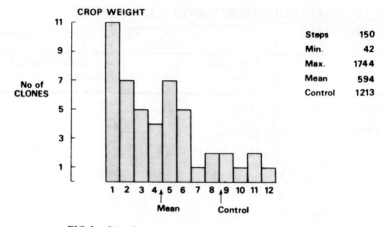

FIG.5. *Distribution of mutant clones for crop weight.*

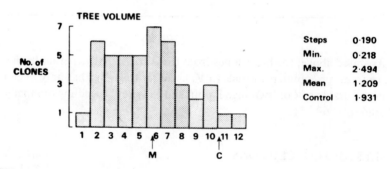

FIG.6. *Distribution of mutant clones for tree volume. M = mean of mutant clones; C = mean of control clones.*

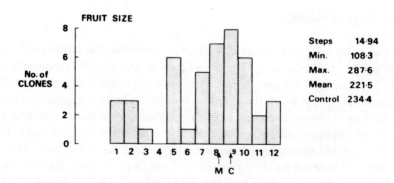

FIG.7. *Distribution of mutant clones for fruit size. M = mean of mutant clones; C = mean of control clones.*

CROP WEIGHT/ TREE VOLUME

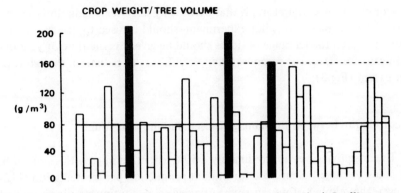

FIG.8. Selected characters of individuals mutant clones of Bramley's Seedling: Crop weight/tree volume. ——— control performance; - - - - - 5% confidence limits of control.

CROP WEIGHT

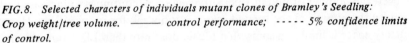
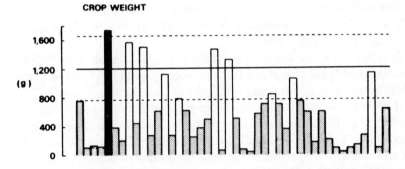

FIG.9. Selected characters of individual mutant clones of Bramley's Seedling: Crop weight. ——— control performance; - - - - - 5% confidence limits of control.

TREE VOLUME

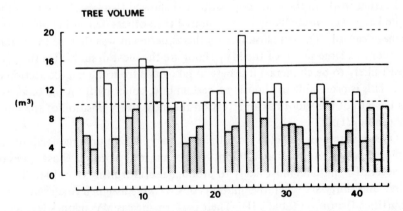
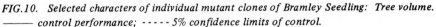

FIG.10. Selected characters of individual mutant clones of Bramley Seedling: Tree volume. ——— control performance; - - - - - 5% confidence limits of control.

Our experience, therefore, is that before any large-scale radiation experiment is carried out, a small-scale pilot experiment should be used to determine the best treatment. Even then a range of doses should be given, as the time of year or season can also effect the response of apple cultivars to radiation, though to a lesser extent (Fig.3).

The correlation between the damage caused and mutants found

The dose that gives approximately a 50% reduction in growth seems to be a good guide to the efficient dose for mutant production per treated scion (Fig.4). However, the number of mutants per surviving scion continues to increase with dose until the growth of the treated scions is too weak for propagation. Hence, if nursery space is limited a somewhat higher dose rate than LD_{50} (growth) might be more efficient in terms of total mutants produced.

The cultivar being treated has a large effect on the number of mutants that will be found from any particular dose, whether measured directly or relative to the damage caused (Table III).

It is possible to use the damage caused to the MV_1 shoots as an indication of the number and types of mutants that may subsequently be expected to show in the MV_2 generation, as was demonstrated in the following experiment [2].

Two hundred and sixty-six irradiated grafts of Bramley's Seedling were graded according to their MV_1 growth habit (all treated with 6–7 krad gamma rays). Tables IV and V show how this affected the number of mutants selected, with the most affected (group 6) giving the greatest proportion of mutants buds per scion propagated. This advantage was not so great when the number of different mutant produced from each scions treated was recorded, but still showed that in a conditio of limited space pre-selection can be of value.

Further work on the same population [3] showed that certain buds on the MV_1 budstick were more likely to be mutated than others depending on the type of damage caused. The conclusion from this experiment was that, as the mutants that occur as a large sector of the MV_1 shoot are the most desirable (as they are the least likely to be chimeras), it is best to take buds from the middle section of a shoot. This is preferable to choosing a certain bud, counting from the base, as the overall length of the shoot has an effect on the most likely position of the mutated tissue (Table VI).

Once mutants are selected from the MV_2 they have to undergo trials of their cropping capacity, etc., and an attempt has been made to predict those types of MV_2 vegetative mutants most likely to produce useful fruiting trees [4]. Two hundred and thirty-six vegetative mutants of Cox were assigned to eleven (some-what artificial) groups (Table VII). Their performance as MV_3 clones was recorded (Table VIII) and those in the less extreme categories were, in the main, the most

TABLE X. HEIGHTS, COEFFICIENT OF VARIATION AND
FEATHERING HABIT OF ONE-YEAR-OLD TREES OF
SELECTED COMPACT CLONES OF BRAMLEY'S SEEDLING ON
MM106 (mean of 50 trees per clone)

Clone No.		Height (% of control)	S.D.	Coefficient of variation (%)	Lateral branches (+ present, − absent)
Uniform clones					
Control		100.00	3.35	3.36	−
20		89.01	2.61	2.94	+
36		81.01	3.44	4.24	−
46		78.53	2.67	3.41	−
58		88.15	3.29	3.73	+
68		81.38	2.73	3.35	−
91		84.09	4.10	4.87	−
94		96.18	4.99	5.19	+
Uneven clones					
10	Tall	99.30	5.18	5.22	−
	Short	73.92	4.76	6.45	−
27	Tall	99.50	3.55	3.57	−
	Short	80.10	5.27	6.57	+
42	Tall	101.00	4.00	3.97	−
	Short	89.50	3.67	4.10	−
44	Tall	102.00	3.86	3.80	−
	Short	87.32	3.41	3.90	−
69	Tall	96.26	5.90	6.12	−
	Short	80.58	3.21	3.98	−

likely to succeed in cropping trials (Table IX), although the differences did not
reach statistical significance.

Finally, as a result of the trials, mutant clones are selected for release to the
industry. Figures 5–10 inclusive are taken from Lacey and Campbell [4].
Figures 5–7 show the range of mutants for the important characters of crop and
tree size. In the case of Bramley's Seedling selection was for crop weight per unit

tree volume in order to produce select trees that would give a high cropping density per hectare, and Fig.8 shows that some of the mutants did exceed the control for this character. The solid horizontal line on Figs 8—10 represent the control level, and the dotted horizontal lines its 5% confidence limits.

The clones selected for crop and quality characters (12 clones in the case of Bramley's Seedling) were multiplied to produce trees for commercial trials. Seven of these clones produced perfectly uniform progeny when propagated by budding (coefficients of variation of 5% or less), but five clones split into two distinct types (Table X). One of these types was, in each case, not significantly different from the control clone, so it would seem that these clones were still chimeras (presumably periclinal) after four vegetative generations since irradiation. The other seven apparently stable clones have been planted in replicated trials at 12 sites in the United Kingdom, where they are showing no signs of instability.

Further multiplication of selected mutant Cox clones is giving the same results, about half appearing stable on large-scale propagation. Some techniques are now being tried to produce stable homohistant forms from the known periclinal chimeras. These include: the forcing of adventitious buds (but with little success for British cultivars), tissue culture (meristem proliferation) and the culture of shoots from roots. Tissue culture is possible with apple cultivars, but it is not yet known if it will succeed in resolving periclinal chimeras. The production of shoots from the roots of apple cultivars results in plants derived solely from the inner tissue layers, and hence produces homohistant plants from periclinal chimeras. Unfortunately it is a long process and depends on the mutant genotype being present in the inner layers of the plant, which may not always be the case.

CONCLUSION

The mutation breeding of apple cultivars can produce useful mutants provided that the populations used are large, and that selection is for a type of plant that is likely to occur as a mutant, e.g. dwarf or compact forms of apple are much more likely to be found than larger than normal forms. Using normal propagation procedures pre-selection at the various stages can be used to increase the efficiency of such a programme, but only at the risk of the loss of some of the less extreme mutants, and unfortunately these are probably those best suited to commercial acceptance. Mutants produced by this technique seem to have about a 50% chance of being homohistant, after four generations of vegetative multiplication, again emphasizing the need for a large programme so that there is still a choice of mutants at this stage. Techniques are becoming available which should help overcome the problem of chimeras.

REFERENCES

[1] LACEY, C.N.D., "Propagation and irradiation techniques used to produce and isolate mutants in fruit trees", The Use of Ionizing Radiation in Agriculture (Proc. E.E.C. Workshop, Wageningen, 1976), Rep. EUR.5815 EN (1977) 493.

[2] LACEY, C.N.D., CAMPBELL, A.I., Character changes in induced mutants of Bramley's Seedling apple, Acta Hortic. 75 (1977) 51.

[3] LACEY, C.N.D., CAMPBELL, A.I., The position of mutated sectors in shoots from irradiated buds of Bramley's Seedling apple, Environ. Exp. Bot. 19 (1979) 145.

[4] LACEY, C.N.D., CAMPBELL, A.I., The characteristics and stability of a range of Cox's Orange Pippin apple mutants showing different growth habits, Euphytica 28 (1979) 119.

Bibliography

LIST OF PAPERS (EXCLUDING ANNUAL REPORTS) PUBLISHED AS A RESULT OF THE MUTATION BREEDING OF APPLES PROJECT OF LONG ASHTON RESEARCH STATION

CAMPBELL, A.I., Clonal variation in 'Cox's Orange Pippin': Fruit Present and Future 2 (1973) 75.

CAMPBELL, A.I., Which Cox clone is best, The Grower 81 (1974) 116.

CAMPBELL, A.I., Compact apple trees produced by irradiation for use in the meadow orchard, Compact Fruit Tree 9 (1976) 43.

CAMPBELL, A.I., "Improved planting material", Horticultural Education Association, Presidential Address, Autumn Conference, Sutton Bonington, England (1976) 151.

CAMPBELL, A.I., "The growth and cropping characteristics of some induced mutants of apple cultivars", Proc. 20th Int. Horticultural Congress, Sydney, Australia (1978) Abstract 1067.

CAMPBELL, A.I., LACEY, C.N.D., Compact mutants of Bramley's Seedling apple, J. Hortic. Sci. 48 (1973) 397.

CAMPBELL, A.I., LACEY, C.N.D., Radiation mutants in apples can produce made to measure trees, The Grower 82 (1974) 1172.

CAMPBELL, A.I., LACEY, C.N.D., "Mutation induction in fruit tree cultivars at Long Ashton", Proc. Eucarpia Fruit Section Symposium 5, Top Fruit Breeding, Canterbury 1973 (1975) 40.

CAMPBELL, A.I., WILSON, D., "Prospects for the development of disease-resistant temperate fruit plants by mutation induction", Induced Mutations Against Plant Diseases (Proc. Symp. Vienna, 1977), IAEA, Vienna (1977) 215.

LACEY, C.N.D., Bad maidens cost more (A study of the effect of the quality planting material), The Grower 85 (1976) 424.

LACEY, C.N.D., "Progress report on techniques used to produce and isolate mutants of fruit trees", Proc. E.S.N.A. Contact Group 6, Caderache, France, 1975 (1976) 26.

LACEY, C.N.D., "Propagation and irradiation techniques used to produce and isolate mutants in fruit trees", The Use of Ionizing Radiation in Agriculture (Proc. E.E.C. Workshop, Wageningen 1976), Rep. EUR. 5815 EN (1977) 493.

LACEY, C.N.D., "The mutation spectrum of Cox's Orange Pippin resulting from gamma irradiation", Proc. 6th Symp. Eucarpia Fruit Section, Wageningen 1976 (1977) 46.

LACEY, C.N.D., CAMPBELL, A.I., Character changes in induced mutants of Bramley's Seedling apple, Acta Hortic. **75** (1977) 51.

LACEY, C.N.D., CAMPBELL, A.I., The characteristics and stability of a range of Cox's Orange Pippin apple mutants showing different growth habits, Euphytica **28** (1979) 119.

LACEY, C.N.D., CAMPBELL, A.I., The position of mutated sectors in shoots from irradiated buds of Bramley's Seedling apple, Environ. Exp. Bot. **19** (1979) 145.

LACEY, C.N.D., CAMPBELL, A.I., "The characters of some selected mutants clones of Bramley's Seedling apple, and their stability during propagation", Proc. Eucarpia Fruit Section Symposium, Angers, France, 1979 (1980) 301.

MUTAGENESIS APPLIED TO IMPROVE FRUIT TREES
Techniques, methods and evaluation of radiation-induced mutations

B. DONINI [†]
Laboratorio Applicazioni in
  Agricoltura del CNEN,
CSN Casaccia, Rome,
Italy

Abstract

MUTAGENESIS APPLIED TO IMPROVE FRUIT TREES: TECHNIQUES, METHODS AND EVALUATION OF RADIATION-INDUCED MUTATIONS.

Improvement of fruit tree cultivars is an urgent need for a modern and industrialized horticulture on which is based the economic importance of many countries. Both the cross breeding and the mutation breeding are regarded as the methods to be used for creating new varieties. Research carried out at the CNEN Agriculture Laboratory on mutagenesis to improve vegetatively propagated plants, under the FAO-IAEA Co-ordinated Research Programme, has dealt with methods of exposure, types of radiations, conditions during and after the irradiation, mechanisms of mutation induction, methodology of isolation of somatic mutations and evaluation of radiation-induced mutations in fruit trees. Problems associated with these aspects have been evaluated, which is very important for the more efficient use of radiation in the mutation breeding. Mutants of agronomical importance (plant size reduction, early ripening, fruit colour change, nectarine fruit, self-thinning fruit) have been isolated in cherry, grape, apple, olive and peach and they are ready to be released.

INTRODUCTION

For a modern and industrialized horticulture new varieties are required. Fruit growers are looking for varieties which provide an economical income and are competitive with other crops. The main attributes of the varieties should be: constant and high productivity, early entering into production, resistance to parasites and disease, all fruit ripening at the same time, easy harvesting by hand or machine, compact growth habit, durability under manipulation and cold storage.

---

* Contribution No. 596 from Divisione Applicazioni Radiazioni CNEN, CSN Casaccia: Research carried out in co-operation with the IAEA under Research Agreement No. 1507.

† Present address: Plant Breeding Section, Joint FAO/IAEA Division of Isotope and Radiation Applications of Atomic Energy for Food and Agricultural Development, International Atomic Energy Agency, Vienna.

Acceptance of new varieties by consumers is based on the high quality of the fruit and its attractiveness, and the ability for it to be marketed at different times of the year.

Mutagenesis may complement conventional breeding methods and has more recently contributed to creating new varieties of agronomical interest.

Since 1972 the Joint FAO/IAEA Division has encouraged a Co-ordinated Research Programme of irradiation applied to the improvement of vegetatively propagated crops and woody perennials which leads to scientific co-operation among the institutes of several countries.

The research carried out in our laboratory covered methods of exposure, types of irradiation, conditions during and after irradiation, mechanisms of mutation induction, methodology of isolation of somatic mutations and evaluation of radiation-induced mutations in fruit trees.

TECHNIQUE OF MUTAGENIC TREATMENT

Both acute X- or gamma-rays and chronic exposures in the gamma-field applied on dormant or actively growing materials have been proved to induce somatic mutations.

It should be emphasized that plants subjected to chronic irradiation may possibly tolerate a higher total dose (10 times higher) than acute irradiation. However, the frequency of somatic mutations obtained in plants under chronic irradiation is not more than that obtained after acute irradiations, due in the former case to the great difficulty of mutation isolation.

For acute irradiation the total doses for fruit species so far investigated, which may be adequately used for mutagenic purposes, range between 2 krad and 7 krad depending on the radiosensitivity of the species.[1] Even the choice of the exposure rate has to be taken into consideration; in fact it has been found that, by increasing the exposure, a higher frequency of somatic mutations may be found in the lower zone of the V_1 shoot.

From our data we consider that exposure rates ranging between 200 rad/h to 2000 rad/h for acute irradiation and between 5 rad/day and 35 rad/day for chronic irradiation over one year may be suitably used.

The increased rate of somatic mutations obtained in cultivars of *Prunus avium, Vitis europea* and *Antirrhinum majus* when a second treatment is applied on the V_1 shoot coming from an irradiated bud confirms this procedure and indicates that it may be put into practice.

---

[1] 1 rad = 1.00×10^{-1} Gy.

In spite of the more uniform response and higher biological efficiency from using densely ionizing radiations (fast and thermal neutrons) compared with X-ray or gamma-ray facilities the latter are largely used as they are more readily available and efficient in practice.

To enhance the frequency of somatic mutations from the results obtained in cherry, apple and peach it seems advisable to irradiate buds that are beginning to grow rather than summer resting buds.

The use of rooted scions or one-year grafted trees may enable an increase to be made in the dose normally used for dormant buds that have to be grafted. The shielding of the basal part of the scions during irradiation give a greater chance of success in grafting.

MECHANISM OF MUTATION INDUCTION

The frequency of somatic mutations, increased by the irradiation of a genetical pre-existing chimeric structure, is rather higher compared with that resulting from a true mutational event.

Both acute and chronic irradiation applied to dormant and growing buds of peach with the epidermal leaf tissues genetically green (G) and the internal tissue yellow (Y), and of *Corylus avellana atropurpurea* with the L-I epidermis genetically green and the inner tissue red, gave rise to a much larger number of mutated sectors.

This shows the irradiation potentiality for the appearance of somatic mutations in material having such structures as a consequence of cell layer rearrangement. The chance of cells shifting from outer to inner layers and vice versa has been clearly observed in peach cytochimeras 4n, 2n, 2n, and 2n, 4n, 4n, which were irradiated with total doses of 300, 400, 900, 1200 and 1500 rad.

Radiation treatments may induce, by chance, a mutational event in the cell(s) belonging to an external or inner cell layer of the shoot meristem. We have evidence in a peach nectarine mutant isolated by irradiation of the cultivar Fertilia, heterozygous for character Gg, that the mutation resides in the epidermal layer (periclinal ectochimera).

In fact the cross of the mutant with the variety Panamint (gg) gives rise to a segregation of 50% producing pubescent or fuzzless fruits respectively. In spur-type mutants isolated in cherry varieties, the mutation induced probably resides in the inner layers of the shoot apex.

The induced mutations, similarly to those spontaneously occurring, may affect monofactorial as well as bi- or polyfactorial characters, and the genetical change is from dominant to recessive. In fact, in the mutants induced in peach, apple and cherry the modified characters were: fruit pubescence (single gene), flowering time and fruit harvest period.

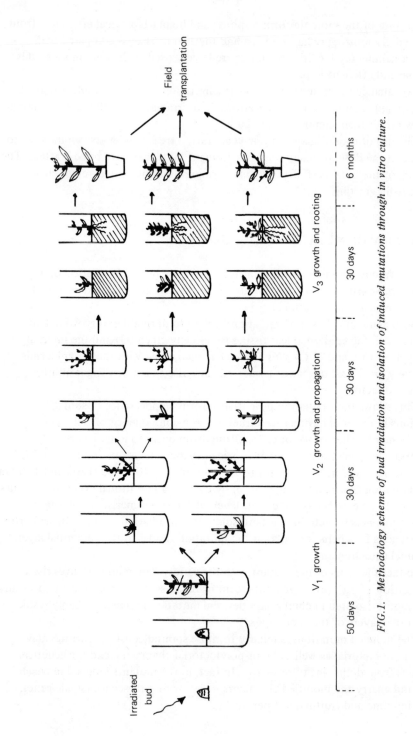

FIG.1. Methodology scheme of bud irradiation and isolation of induced mutations through in vitro culture.

METHODOLOGY OF SOMATIC MUTATION ISOLATION

The irradiation of an organized and multicellular structure such as a bud, which is the material more frequently used for the fruit species treatment, may induce mutational events in the cell(s) of the different meristematic territories already present in the shoot apex which leads to a sectorial or mericlinal chimeric situation.

To isolate the induced mutations in a periclinal and constant form, techniques developed in the different fruit species investigated aim to promote the V_1 growth from the treated bud and to force, through clonal propagation, the development of the buds along the V_1 shoots into V_2 shoots. The results obtained in cherry varieties, cherry rootstocks $F_{1/12}$ and olive reveal a higher mutation frequency in the V_2 and V_3 shoots originating from the V_1 buds already present as primordia immediately below the apical meristem of the irradiated bud, compared with that from the V_2 and V_3 shoots coming from the V_1 buds of the upper V_1 shoot zone (neoformation region). Thus an histological analysis, at the stage of bud irradiation, may enable the average number of primordia present in the shoot apex to be ascertained for each species, and consequently to indicate the number of V_1 buds that are worthwhile being multiplied.

To reduce the time elapsing between bud irradiation and V_2 shoot growth an in vitro technique of handling the material has been developed in the apple cultivar Annurca (Fig. 1), which may be applied for other crops.

Through this procedure there can be continuous growth and micropropagation of the material until the V_2 in four months under controlled environmental conditions, which may facilitate the isolation of the induced somatic mutations and enhance their frequency.

The V_2 shoots can be rooted or grafted on seedling material giving rise in six months to V_3 plantlets ready to be planted in the field.

GENETICAL AND AGRONOMICAL EVALUATION OF RADIATION-INDUCED MUTATIONS IN FRUIT TREES

Within the cultivars of each species investigated, cherry, grape, peach and olive, subjected to the same total dose, differences in terms of mutation frequency have been observed.

They are very relevant in cherry: D. di Vignola II 8.5‰, D. di Vignola I 7.3‰, B. Napoléon 6.1‰, B. Moreau 4.7‰, B. Burlat 1.2‰; and in olive: Ascolana 14.1‰, Moraiolo 4.2‰. Even in peach and grape a higher mutation frequency has been observed in Fertilia compared with Favorita (peach), and in Bonarda and Dolcetto compared with Freisa and Delight grape cultivars.

TABLE I. MUTATIONS INDUCED BY IRRADIATION IN FRUIT TREES
Joint projects of the Agricultural Laboratory of CNEN and the Horticulture Institute of MAF, CNR and Rome University

Species	Cultivar	Mutations recovered (No.)	Characters
PEACH	Fertilia	4	Nectarine, early ripening fruit
	Favorita	1	Narrow leaf and glabrous fruit
CHERRY	Bigarreau Moreau	3	Spur type, early entering into
	B. Napoléon	3	production, early production per tree
	B. Burlat	1	” ” ”
	D. di Vignola I	4	” ” ”
	D. di Vignola II	17	” ” ”
	Mora di Vignola	5	” ” ”
	Mora di Cazzano	6	” ” ”
	F 12/1 (rootstock)	4	Spur type, rooting ability
GRAPE	Bonarda	2	Short internode, thinning grape
	Dolcetto	2	Early ripening fruit, resistance to berry fall
	Regina Vigneti	1	Small grape and seedless
OLIVE	Ascolana	Several mutations	Spur type, early entering into production, higher oil content
	Moraiolo	” ”	” ” ”
	Frantoio	” ”	” ” ”
	Leccino	” ”	” ” ”
APPLE	Annurca	Several mutations	Spur type, wider angle of shoot insertion, red fruit skin

The genetical background and the different degrees of heterozygosity of the cultivars may explain the results obtained.

Differences among the V_3 plants arising from the propagation of a mutated V_2 clone have been frequently observed in the plant 'habitus' (spur to normal plant), in the flowering period (10 days before to 10 days after the control), in pollen abortion and in fruit size (very small to higher than control). It might be argued that in such cases the V_2 shoots still have sectorial or mericlinal chimera. For this reason it is advisable before practical use is made of the isolated mutation to ascertain the genetical stability reached only through propagation in a periclinal chimera structure. The mutant induced in peach for the nectarine character of the fruit, which has been largely vegetatively propagated for agronomical purposes, confirms in its descendants the uniform stability and the maintenance of the somatic mutation.

The mutations so far isolated in the different fruit species irradiated are reported in Table I.

In the joint projects of the Agriculture Laboratory of CNEN and the Horticulture Institutes of the Ministry of Agriculture, the National Research Council and Rome University, several agronomical trails are now in progress to evaluate the mutants in different regions of Italy. A few mutated cherry clones for 'compact habitus', which have already been evaluated, are now distributed for planting on a large scale.

BIBLIOGRAPHY

DONINI, B., "Induction and isolation of somatic mutations in vegetatively propagated plants", Improvement of Vegetatively Propagated Crops and Tree Crops through Induced Mutations (Proc. First Research Co-ordination Meeting, Tokai, 30 September – 4 October 1974), IAEA-173 (1975) 35.

DONINI, B., "Breeding methods and applied mutagenesis in fruit plants", Proc. Workshop on the Use of Ionizing Radiation in Agriculture, Wageningen, 22–24 March 1976, Rep. E.U.R. 5815 EN (1977) 453.

DONINI, B., "The use of radiations to induce useful mutations in fruit trees", Improvement of Vegetatively Propagated Crops and Tree Crops through Induced Mutations (Proc. Second Research Co-ordination Meeting, Wageningen, 17–21 May 1976), IAEA-194 (1976) 55.

DONINI, B., "Evaluation of the mutants induced through radiations in different fruit species" Improvement of Vegetatively Propagated Crops and Tree Crops through Induced Mutations (Third Research Co-ordination Meeting, Skierniewice, 22–26 May 1978) (Unpublished).

DONINI, B., "Induced mutations and breeding methods in fruit plants", XXth International Horticultural Congress, Sydney, 15–23 August 1978.

DONINI, B., Evaluation of spur mutants of sweet cherry, Mutation Breeding Newsletter, No. 11 (Feb. 1978) 8.

DONINI, B., Evaluation of radiation induced mutants in fruit trees, Monografie di Genetica Agraria **IV** (1979) 313.

DONINI, B., CRESTI, M., CIAMPOLINI, F., PACINI, E., "Studio ultrastrutturale dell'effetto delle radiazioni gamma sul tessuto trasmittente stilare di ciliegio *(Prunus avium)*", Seminario della Società Orticola Italiana: La fertilità nelle piante da frutto, Bologna, 15 December, 1978, 577.

DONINI, B., PETRUCCIOLI, G., ROSELLI, G., Impiego delle radiazioni per indurre mutazioni in piante di olivo, Ann. Istituto Sperimentale Olivicoltura **3**, Cosenza (1975).

DONINI, B., ROSELLI, G., "The use of mutation breeding to obtain nectarines", Riv. Ortoflorofruticoltura Italiana **61** 3 (1977) 175.

DONINI, B., VEGLIO, P., "Impiego della mutagenesi per il miglioramento di studio sull'uso di tecniche nucleari per il miglioramento e la difesa dei fruttiferi, CSN Casaccia, 8–10 April 1974, **1** (1976) 13.

FIDEGHELLI, C., ROSATI, P., DONINI, B., "Mutanti compatti indotti con radiazioni ionizzanti in varietà di ciliegio", Giornate di studio sull'uso di tecniche nucleari per il miglioramento e la difesa dei fruttiferi, CSN Casaccia, 8–10 April 1974 **1** (1976) 57.

PETRUCCIOLI, G., FILIPPUCCI, B., DONINI, B., "Impiego della mutagenesi per l'ottenimento di forme nanizzate nell'olivo", Giornate di studio sull'uso delle tecniche nucleari per il miglioramento e la difesa dei fruttiferi, CSN Casaccia, 8–10 April 1974 **1** (1976) 95.

ROMISONDO, P., DONINI, B., "Induzione di mutazioni gemmarie nella vite mediante radiazioni ionizzanti", Giornate di studio sull'uso di tecniche nucleari per il miglioramento e la difesa dei fruttiferi, CSN Casaccia, 8–10 April 1974 **1** (1976) 85.

ROSELLI, G., DONINI, B., "Mutazioni radioindotte in ciliegio, pesco ed olivo", Giornate di studio sull'uso delle tecniche nucleari per il miglioramento e la difesa dei fruttiferi, CSN Casaccia, 8–10 April 1974, **1** (1976) 71.

DEVELOPMENT OF COMPACT MUTANTS IN APPLE AND SOUR CHERRY*

S.W. ZAGAJA, A. PRZYBYŁA, B. MACHNIK
Research Institute of Pomology
    and Floriculture,
Skierniewice,
Poland

Abstract

DEVELOPMENT OF COMPACT MUTANTS IN APPLE AND SOUR CHERRY.

During the period 1973 — 79 studies were conducted with the aim of developing compact mutants in apple and cherry cultivars and in apple vegetative rootstocks. During the investigations the effect of the dose of gamma rays on frequency of the mutants was studied. Attempts were also made to evolve a micropropagation technique adapted to propagate P 2 and P 22 apple rootstocks, as an aid in mutation breeding. Several mutants were produced in all the material studied, but none of them have yet reached a sufficient developmental stage to enable their complete assessment. On the basis of the results obtained so far the following conclusions can be drawn: higher doses of irradiation resulted in higher frequency of mutants in most apple cultivars and apple rootstocks; in sour cherries the effect of dose depended on the cultivars. Among V_1 shoots developed from sleeping buds on irradiated scion wood, compact mutants were found; their frequency, however, was about 60% lower than among V_1 shoots developed directly from irradiated dormant buds. In apple rootstocks A 2 and M 26 several dwarfed mutants were found; some of these produced thorny plants and some had lower rooting ability; both these characteristics are inferior from the practical point of view. Multiplication and rooting media for in vitro propagation of apple rootstocks, worked out for M 26, were found unsuitable for the rootstocks P 2 and P 22; modifications made in the growth substance composition of the above media enabled satisfactory propagation to be obtained.

INTRODUCTION

Cross breeding, based on crossing progenitors each having different desirable traits, attempts to synthesize new genotypes combining traits from both parents. As a result new cultivars are produced which in most cases differ from both parents with respect to several characteristics.

Mutation breeding of vegetatively propagated crops aims at improving a given variety in a limited number of aspects. Judging from the occurrence in nature of these types of mutants, mutation breeding of fruit crops appears feasible [1]. There are also reports on successful improvement of fruit plants through mutation breeding [2].

---

* Research supported by the IAEA under Research Contract No. 1501.

AIMS

The main aims of the studies conducted during the reporting period were as follows.

The practical aims were to develop compact-type mutants in apple, sour cherry and pear cultivars as well as dwarf apple rootstocks of commercial importance in the Polish fruit industry.

The scientific aims were:

1. To find out the effect of the dose rate of gamma rays on the incidence of compact-type mutants
2. To study the incidence of compact-type mutants obtained from sleeping buds developed on irradiated shoots
3. To work out a technique of mass production of apple rootstocks from tip culture as an aid in mutation breeding.

MATERIAL AND METHODS

Plant material

Leaf buds of dormant vegetative shoots were used to irradiate fruit tree cultivars. In vegetative apple rootstocks leaf buds of dormant one-year-old rooted plants were used.

Three- to five-mm long tips taken from freshly developing shoots (not exceeding 50 mm in length) were used for micropropagation.

Irradiation

Gamma rays from a ^{60}Co source were used in all experiments. Acute irradiation was performed at the total doses indicated in Table I.

Median parts of irradiated shoots were cut into three-bud scions and were grafted on to rootstocks or top worked to mature apple trees.

Irradiated rootstocks were lined into a fruit tree nursery using standard planting distances.

Selection techniques

At the end of the first growing season all the shoots developed from irradiated buds (V_1) were collected and compared with the control shoots, i.e. not the treated ones, of the respective cultivars with a comparable diameter at the base.

The harvested shoots, with clearly shortened internodes, were grafted on to rootstocks to produce V_2 shoots. Standard cultivars were included randomly

TABLE I. INCIDENCE OF MUTANTS DEPENDING ON THE IRRADIATION DOSE

Cultivar	Dose (krad)[a]	Number of V_1 shoots	Number of mutants obtained	Percentage of mutants
Experiment I (1972) Apple cultivars				
Bancroft	2.5	2018	nil	0
	5.0	2626	2	0.07
Cox's Orange	2.5	2071	4	0.20
	5.0	1852	3	0.16
Spartan	2.5	814	2	0.24
	5.0	451	2	0.43
Experiment II (1974) Apple cultivars				
Bancroft	3.0	1971	2	0.10
	4.0	2646	7	0.26
	5.0	1820	10	0.54
Close	3.0	1176	4	0.34
	4.0	1179	4	0.34
	5.0	835	3	0.34
Experiment III (1973) Cherry				
Körözer	2.0	664	5	0.75
	3.0	480	1	0.20
	5.0	627	1	0.15
Nefris	2.0	1575	1	0.06
	3.0	1534	2	0.13
	5.0	1290	2	0.15
Schattenmorelle	2.0	1540	9	0.59
	3.0	1659	10	0.60
	5.0	1641	6	0.36
Experiment IV (1973) Apple rootstocks				
A2	2.5	374	2	0.53
	3.5	241	6	2.48
M 7	2.5	305	nil	0
	3.5	343	nil	0
M 26	2.5	253	1	0.20
	3.5	132	3	2.72
MM 106	2.5	246	nil	0
	3.5	246	nil	0

[a] 1 rad = 1.00×10^{-2} Gy.

among the tested ones. The height of the maidens and the length of their internodes were recorded in August of the same year. The apparently unchanged
maidens were discarded and the compacts were repropagated once more in the
nursery. Uniformly appearing compact-type V_3 maidens were planted in a test
orchard to be brought to fruiting. The remaining ones were discarded.

Dwarf-appearing V_3 rootstock mutants were layered next spring to produce
rooted plants. These were lined in the nursery to be budded with commercial
cultivars in order to produce trees for further studies in a test orchard.

Early in the second spring after grafting, the original irradiated grafts were
cleared of all buds in order to force the growth of new shoots from sleeping buds.
The new shoots developed during growing season were harvested in December and
treated in the same manner as those described above.

In the test orchards vigour of growth of the trees, spur formation, flower bud
formation, flower fertility, crop size, fruit size, shape and colour were assessed.

Micropropagation technique

A micropropagation technique developed by Jones and co-workers [3] was
initiated to propagate two vegetative apple rootstocks, namely P 2 and P 22, and
yielded negative results. It was found after several attempts that the composition
of the media developed by these authors for M 26 rootstock was not suitable for
the genotypes used in our studies. Therefore the work concentrated mainly on
developing the proper media. The outcome is reported in the section on results.
It is planned that after further refinement of the media irradiation of tips will be
started.

RESULTS AND DISCUSSION

Mutants selection

The total number of dormant buds irradiated during the reporting period
and the total number of V_1 shoots are presented in Table II. The mortality of
irradiated buds in apples never exceeded 10%. In sour cherries it was higher,
ranging from 9 to 34%, depending on the cultivar. Mortality of the buds was
not related to the total dose of irradiation. This was probabaly because the total
doses used were relatively low, ranging for apple cultivars and apple rootstocks
between 2 and 5 krad, and for cherries between 2 and 4 krad. The total number of
mutants derived directly from irradiated buds is also shown in Table II.

The total number of shoots developed from sleeping buds was two to three
times higher than that of V_1 shoots originating directly from irradiated buds.
Among the shoots developed from sleeping buds the following numbers of

TABLE II. TOTAL NUMBER OF IRRADIATED BUDS, V_1 SHOOTS
PRODUCED AND COMPACT MUTANTS OBTAINED

Plant material	Number of irradiated dormant buds	Number of V_1 shoots	Number of preliminary selected V_2 shoots	Number of mutants
Apple cultivars	12450[a]	10200	303	27
Apple rootstocks	2400[a]	2010	107	12
Pear cultivars	750[a]	660	0	–
Sour cherry cultivars	4500[a]	2563	29	9

[a] Buds producing rosettes of leaves only were considered as survivals.

compact mutants were found: apple cultivars, 26; apple rootstocks, 12; cherry
cultivars, 28; pear cultivars, nil. Therefore it can be concluded that the incidence
of compact mutants in apple cultivars and in apple rootstocks was higher within
the shoots developed directly from irradiated buds than from the sleeping ones.
In the case of sour cherries the incidence of mutants in both groups of shoots
was similar. The total number of mutants obtained up to date is as follows:
apple cultivars, 53; apple rootstocks, 23; sour cherry cultivars, 37.

From the methodological point of view it seemed interesting to ascertain
the effectiveness of the method of mutant selection used in these experiments.
It was found that from all the originally selected compact-type shoots (V_1) only
about 10% proved to be stable mutants. A similar ratio was found in shoots
developed from sleeping buds.

Mutant incidence as affected by the dose rate

There were four experiments in which differentiated doses of irradiation were
used. The details of the experiments and the results obtained are presented in
Table I.

As can be seen from the data in this table, an increase in dose rate resulted
in most cases in a higher incidence of compact mutants. A clearly opposite
tendency was noted once only for the Körözer sour cherry cultivar.

With the exception of this last case, the results obtained in our experiments
are in agreement with those reported by several authors [4].

The total number of mutants per cultivar obtained in these studies was
lower than those reported by some other workers on fruit plants [5, 6].

TABLE III. GROWTH, FLOWERING AND FRUITING OF APPLE MUTANTS PLANTED IN 1973

Cultivars and clones	Trunk diameter 1979	Total length of 1-year-old shoots in two years (cm/tree)	Number of 1-year-old shoots in two-year-old trees	Length of internodes in two-year-old trees (average) (cm)	Number of flower buds per tree in the first year of flowering	Total yield of fruits in three years of fruiting (kg/tree)
Boscoop standard	9.1	2206.6	98.9	2.5	11.3	21.3
Boscoop S1	6.7	996.1	39.1	1.7	31.8	–
Boscoop S2	7.3	1541.0	44.8	2.0	14.6	46.8
Macoun standard	8.6	1954.4	68.1	2.3	24.9	31.6
Macoun S1	7.3	863.5	31.6	1.8	23.3	18.7
Macoun S2	7.3	1027.0	44.0	1.8	1.0	13.3
McIntosh standard	8.2	2211.2	88.0	2.3	80.6	32.3
McIntosh E	8.4	1138.4	33.4	1.9	104.0	56.4
McIntosh 1 A	7.6	1057.4	28.4	1.7	12.2	11.7
McIntosh 1 B	6.7	713.1	21.9	1.7	2.9	6.4
McIntosh 1 C	7.3	875.4	28.8	1.7	1.4	8.2
McIntosh 2 A	6.9	900.5	27.6	1.7	11.5	9.9
McIntosh 2 B	8.2	1345.3	35.9	1.7	23.5	15.8

Two complementary factors seem to be responsible for these differences. First, the irradiation doses used in our experiments were lower than those used by these authors. Second, the selection criteria used in our experiments were very rigid; this might have led to discarding mutants which could be classified as semi-compact ones.

We are aware of the fact that, in order to speculate scientifically on the mutability of different cultivars, a large amount of plant material is needed. Nevertheless our results seem to support the opinions expressed by earlier authors that different genotypes of plants of the same species differ in their mutability [4].

Some pomological characteristics of the mutants obtained

It will take another five to six years' study before all the pomological characteristics of all the selected mutants will be available. At present only preliminary data are available.

Apple cultivars

There are nine mutants, developed under this project, which have already reached bearing age, i.e. two mutants of Belle de Boscoop, two mutants of Macoun and five mutants of McIntosh cultivar. All of them have the features of compact-type trees (Table III).

With the exception of Belle de Boscoop S 2 and Macoun S 1 the remaining mutants are inferior in their productivity compared with the standard cultivars. This is not unexpected because similar data have been reported elsewhere [7, 8].

There is another group of four apple mutants planted in the orchard in 1975. They came into first bearing in 1979. Mutants Boiken S 13 and McIntosh 8/XLII/74 gave higher yields than the standards.

The remaining apple mutants were planted in the trial orchard in the years 1976 to 1979 and no data on their fruiting are yet available.

Cherries

So far only the data on vegetative growth of sour cherry mutants are available (Table IV).

The most striking features of all cherry mutants are shortened internodes and reduced number of branches. Similar type of growth has been reported also for apple induced and spontaneous mutants [7].

Apple rootstocks

Out of a total of 23 rootstock mutants obtained, 13 were discarded at the V_3 stage because they were producing very thorny plants, which is a serious

TABLE IV. GROWTH VIGOUR OF SOUR CHERRY MUTANTS (cm)

Cultivers and clones	Trunk diameter 1979	Total length of shoots	Number of shoots	Average length of shoots	Length of internodes
Körözer standard	2.6	528.2	41.0	12.8	1.6
Körözer 6	1.8	360.9	26.4	13.8	1.4
Körözer 50	1.9	300.0	16.3	18.7	1.5
Nefris standard	4.3	3646.0	206.8	17.6	2.0
Nefris 92	2.7	823.0	54.0	15.2	1.6
Schattenmorelle standard	2.4	1246.7	78.6	17.0	1.7
Schattenmorelle 18 A	1.3	297.8	17.5	17.4	1.4
Schattenmorelle 46	1.3	258.6	12.5	21.5	1.4
Schattenmorelle 56	1.4	265.8	17.5	15.6	1.3
Schattenmorelle 59	1.3	268.5	22.0	12.2	1.2
Schattenmorelle 64	1.2	221.1	14.4	15.6	1.5
Schattenmorelle 67	1.8	339.8	26.0	15.4	1.4
Schattenmorelle 75	1.5	418.5	24.7	17.1	1.8
Schattenmorelle 80	1.5	355.1	21.8	16.1	1.6
Schattenmorelle 81	1.2	214.6	17.0	12.6	1.3

disadvantage from the practical point of view. The mutants retained show different degrees of dwarfness (Table V). From the Alnarp 2 mutants the most dwarfed is clone A 2 12.

All the mutants are characterized not only by shortened total height of the shoots and shorter internodes, but also by reduced diameter of the maidens. The plants grown in the stool beds are still very young so that it is not yet possible to assess the reproduction index of the mutants.

The ability to form adventitious roots is an important feature of vegetatively propagated rootstocks. Preliminary assessment indicates that the rooting ability of none of the M 26 mutants has been changed. Two of the A 2 mutants seem to have considerably lowered rooting ability.

The crucial question concerning the rootstock mutants is whether they will exert a dwarfing effect on the cultivars grafted on to them. An answer to this question will be obtained in the future, when the grafted trees reach bearing age.

TABLE V. GROWTH AND ROOTING OF APPLE MUTANT ROOTSTOCKS (mm)

Rootstocks and clones	Dose of gamma rays (krad)[a]	Height of maidens	Length of internodes	Diameter at the base	Rooting ability[b]
Alnarp 2 standard	–	1038	19	14	4
Alnarp 2 3	2.5	630	12	11	3
Alnarp 2 5	2.5	607	19	10	–
Alnarp 2 11	3.5	390	11	9	4
Alnarp 2 12	3.5	760	15	11	2
Alnarp 2 14	2.5	417	11	11	2
Alnarp 2 16	3.5	607	14	11	3
Alnarp 2 40	2.5	840	18	15	3
Alnarp 2 61	2.5	440	11	14	4
M 26 standard	–	1159	23	17	3
M 26 25	2.5	570	13	14	3
M 26 28	3.5	813	20	16	3
M 26 29	3.5	760	20	16	3

[a] 1 rad = 1.00×10^{-2} Gy.
[b] 5 – very good; 4 – good; 3 – satisfactory; 2 – poor, 1 – nil.

46 ZAGAJA et al.

Micropropagation of apple vegetative rootstocks

As shown by Jones and co-workers [3], using micropropagation techniques, fruit plant material can be propagated very efficiently. It was assumed that the shoots tips, being a starting point in that method of propagation, would be very convenient material for irradiation and for selecting solid mutants in apple rootstocks.

Plant material used in our studies were two dwarfing apple rootstocks, namely P 2 and P 22, developed by Zagaja [9]. The main disadvantage of these otherwise very promising rootstocks is their relatively low propagation index and not fully satisfactory rooting ability.

The first attempts to propagate the rootstocks using the method described by Jones and co-workers [3] failed. After several attempts it was found that the genotypes studied by us require a modified growth medium. Within the last few months the following media have given satisfactory results, although the propagation index is still lower than that reported by Jones for M 26 rootstock:

Multiplication medium [10]		*Rooting medium*	
Agar	0.75%	Perlite	
Saccharose	30 g/l	Saccharose 20 g/l	
Inositol	100 mg/l	Vitamins $B_1 B_2 B_3 B_6$	
BA	1.5 to 2.0 mg/l		0.5 mg/l each
NAA	0.05 mg/l	PG	162 mg/l
		IBA	1 mg/l

It is expected that within the next three to four months the technique will be further refined, and then it is planned to start irradiation experiments.

REFERENCES

[1] LAPINS, K.O., FISHER, D.V., Four natural spurtype mutants of McIntosh apple, Can. J. Plant Sci. **54** (1974) 359.
[2] CAMPBELL, A.J., LACEY, C.N.D., Compact mutants of Bramley's Seeding apple induced by gamma radiation, J. Hortic. Sci. **48** (1973) 397.
[3] JONES, O.P., HOPGOOD, M.E., O'FARRELL, D., Propagation in vitro of M 26 apple rootstock, J. Hortic. Sci. 52 (1977) 235.
[4] INTERNATIONAL ATOMIC ENERGY AGENCY, Manual on Mutation Breeding, 2nd ed., Technical Reports Series No.119, IAEA, Vienna (1977).
[5] LAPINS, K.O., "Induced mutations in fruit trees", Induced Mutations in Vegetatively Propagated Plants (Proc. Panel Vienna, 1972), IAEA, Vienna (1973) 1.

[6] VISSER, T., "Methods and results of mutation breeding in deciduous fruits, with special reference to the induction of compact and fruit mutations in apple", Induced Mutations in Vegetatively Propagated Plants (Proc. Panel Vienna, 1972), IAEA, Vienna (1973) 21.
[7] LAPINS, K.O., Spur types of apple and cherry produced by ionizing radiation, Wash. State Hortic. Assoc. Proc. **59** (1963) 93.
[8] IKEDA, F., Induced mutation in dwarf habits of apple trees by gamma rays and its evaluation in practical uses, Gamma Field Symp. **16** (1977) 63.
[9] ZAGAJA, S.W., "Breeding cold hardy fruits. Proc. XIXth Int. Hortic. Congress. **3** (1974) 9.
[10] MURASHIGE, T., SKOOG, F., A revised medium for rapid growth and bioassays with tobacco cultures, Physiol. Plant. **15** (1962) 473.

SOME MORPHOLOGICAL AND PHYSIOLOGICAL FEATURES OF SEVERAL DWARF APPLE SPORTS INDUCED BY GAMMA RAYS*

F. IKEDA
Institute of Radiation Breeding, NIAS, MAFF,
Ohmiya, Ibaraki-ken

T. NISHIDA
Fruit Tree Research Station, MAFF,
Yatabe, Ibaraki-ken,
Japan

Abstract

SOME MORPHOLOGICAL AND PHYSIOLOGICAL FEATURES OF SEVERAL DWARF
APPLE SPORTS INDUCED BY GAMMA RAYS.

Various types of apple mutants have been induced by gamma-ray irradiation since 1962. Using several clones which had attained fruiting in 1975, investigation of the fruit quality was carried out. Desirable mutants were mostly obtained from chronic treatment. As a result of observations of pollen grains, pollen sterility was detected in almost all clones which originated from acute treatment. The characteristics of several mutants with weak growth habit were assessed for their morphological and anatomical characters. Spur-type mutants were characterized by a lesser formation of secondary wood and slightly longer fibre cells in their wood compared with those of the original cultivars. Also, dwarf-type mutants were characterized by weak secondary growth and by shorter fibre cells in the wood. Various morphological mutants were examined by anatomical methods if they arose as peripheral polyploidal chimeras. In comparison with the nuclear volume of each histogenic layer, almost all the original varieties and mutants show a larger nuclear volume in the subepidermal layer, L-2, than in the epidermal layer, L-1. However, a reverse situation was recognized in clone 500−18 which seems to be a polyploidal chimera such as 4-2-2. Clonal differences in the induction of a brown coloured substance developed in fresh leaves with mono-iodoacetic acid were compared with those of their original cultivars. Most mutants with weak growth habit showed a darker colour reaction than that of the original cultivar. Therefore, the colour reaction might be used as an early detection method of a mutation by a chemical procedure in apple.

1. INTRODUCTION

Various types of apple sports have arisen spontaneously [1−3] and have been induced following irradiation [4−13]. In spite of extensive research carried out in many countries on the induction of mutation in fruit trees, the number of new varieties obtained by irradiation is very low [14, 15]. In apple cultivation,

---

* Research supported by the IAEA under Research Contract No. 1336.

49

TABLE I. THE ORIGIN AND VEGETATIVE CHARACTERISTICS OF
INDUCED COMPACT APPLE MUTANTS AS OF 1974

IRB Clone	Original cultivar	Irradiation method	Dosage (roentgen/day)	Year	Growth habit
500–0	Iwai	Semi-acute	10.0	1968	Dwarf
500–1	Iwai	Chronic	22.9	1968	Spur
500–2	Ralls	Semi-acute	11.4	1964	Spur
500–3	Ralls	Chronic	3.4	1964	Dwarf
500–5	Fuji	Chronic	4.2	1964	Dwarf
500–8	Fuji	Semi-acute	10.0	1969	Dwarf
500–9	Fuji	Semi-acute	10.0	1969	Dwarf
500–10	Fuji	Semi-acute	10.0	1969	Dwarf
500–11	Fuji	Semi-acute	10.0	1969	Dwarf
500–12	Fuji	Semi-acute	10.0	1969	Dwarf
500–13	Fuji	Semi-acute	10.0	1969	Spur
500–14	Fuji	Semi-acute	10.0	1969	Dwarf
500–15	Fuji	Semi-acute	10.0	1968	Dwarf
500–16	Fuji	Short-term chronic	14.1	1969	Sect. chimera
500–17	Fuji	Short-term chronic	28.2	1969	Semi-dwarf
500–18	Fuji	Short-term chronic	28.2	1969	Spur
500–19	Fuji	Short-term chronic	28.2	1969	Semi-dwarf
500–20	Fuji	Semi-acute	10.0	1969	Dwarf
500–21	Fuji	Semi-acute	10.0	1969	Dwarf
500–23	Fuji	Short-term chronic	14.1	1969	Spur
500–24	Fuji	Short-term chronic	14.1	1973	Spur
500–25	Fuji	Short-term chronic	14.1	1973	Dwarf

bud sports which have originated spontaneously are being used for high density
planting with a scion of a spur-type and a dwarfing rootstock. The compact nature
of tree growth is useful not only because of lower harvest costs, but also promising
early productivity. Therefore, mutation breeding for compact growth habits is
significant for apples, especially Fuji which was bred by crossing Ralls and
Delicious in 1939 and which is the preferred cultivar used in Japanese apple
cultivation [16].

In 1962, some experiments with gamma-ray treatment on apple trees including the cultivar Fuji were started at the Institute of Radiation Breeding (IRB) in order to induce some useful mutants such as many spontaneous spur-types of Delicious without changing any of the valuable characteristics. Research results carried out since 1975 on the following five topics are described: (1) Fruit characteristics of mutant clones; (2) Pollen viability of mutant clones; (3) Characteristics of secondary growth in mutant clones; (4) Characteristics of shoot apices among induced mutants; and (5) Colour reaction with iodoacetic acid in leaves. Also, effective methods for obtaining useful mutant clones are discussed.

2. FRUIT CHARACTERISTICS

2.1. Materials and methods

In 1975, several clones which were selected following change of vegetative organs had attained fruiting. Among them, fruits from six sports of the cultivar Fuji and one sport of the cultivar Ralls were analysed. Their origin and vegetative characteristics are represented briefly in Table I. At IRB, the irradiation methods for fruit tree are referred to (Fig. 1) as acute, semi-acute, short-term chronic and chronic irradiation, corresponding to their exposure duration and gamma-ray exposure rate. In chronic irradiation, the tree was treated in a gamma-field (^{60}Co source, about 3000 Ci) and with another method, irradiation treatments were performed in a gamma-room (^{60}Co source, about 1000 Ci) belonging to IRB.[1]

In 1969, the scions (V_1) of bud sports which had been selected up to that time were grafted on the Maruba-kaido (*Malus prunifolia* Boukhausen var. ringo Asami) rootstock and in the next year, nursery trees (V_2) were planted in large pots for forcing fruiting. Data on fruit characteristics of each clone were summarized and analysed in 1974 and 1975. On November 10th, 1974 and 1975, a suitable picking date for the cultivar Fuji, all the fruit was harvested and 10 typical fruits from each clone were analysed.

Fruit shape was measured with a calliper and a measuring tape for the girth of the fruit and both the shortest height and the longest height of each fruit. Values of fruit shape were taken by dividing the girth by the height and the degree of misshapenness or lopsidedness of the fruit was determined by dividing the shortest height by the longest height of each fruit. External colour was measured with a colour difference colorimetric meter (Nihon Denshoku Kogyo Co.) with the value of L, aL, bL. The firmness of fruit flesh was determined with Kiya's penetrator (Kiya Kogyo Co.). Sweetness was determined with a refractometer and acidity was determined by measuring the pH of the apple juice which was extracted with a juicer and filtered.

---

[1] 1 Ci = 3.70 X 10^{10} Bq.

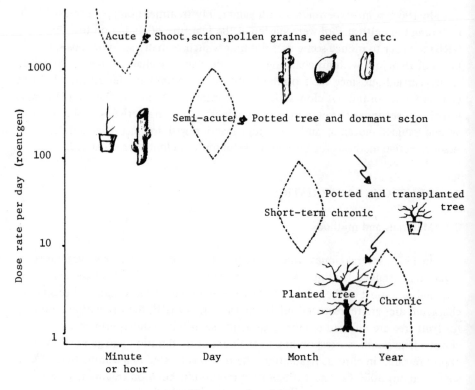

FIG.1. *Schematic presentation of irradiation treatment methods for fruit trees with special reference to dosage rate and exposure duration.*

2.2. Results and discussion

The amount of red colour in the skin differed between clones. The clones which showed a much increased colour were 500−18 and 500−13. Although the fruit of both sports were markedly reddish in appearance, 500−18 showed solid colouring and 500−13 showed broadly striped colouring. On the other hand, 500−23 showed solid colouring but the amount of red colour was less than that of the control (Fig.2). Other clones, 500−17, 500−19 and 500−16, were a little more reddish than the control, but showed similarly striped colouring.

Differences in fruit weight compared with the original were significant in 500−2, 500−13, and 500−23. In particular, fruits of 500−2 were much smaller than the control notwithstanding that its leaf area was the same as that of the control. Also, 500−23 showed differences in fruit shape and seed contents. Its shape was irregular or lopsided, and sometimes was collapsed at the base of fruit

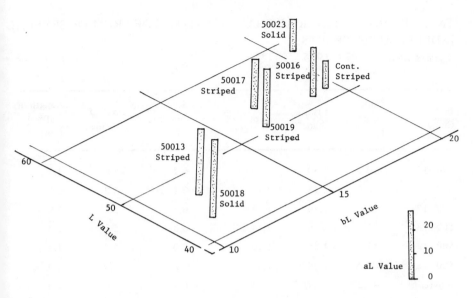

FIG.2. Differences in intensity of red skin colour of the induced mutants and the original cultivar Fuji. Values of L, aL and bL are measured with the colour and colour difference meter.

and also furrowing was present. Moreover, the number of seeds per fruit was much less than the control. The occurrence of a long stalk was more common in clone 500–16 (Table II).

A significant difference in the firmness of the flesh was seen in 500–23 and 500–2. These clones have a firmer texture and contained more starch in the flesh. On the other hand, clone 500–16 showed a little softer texture than the control and was more juicy. The taste of this clone was superior to that of the control. The degree of juice browning was apparently pale. Clone 500–13, which was a broadly coloured type, was inferior in taste and its Brix value was lower than that of the control. The intensity of juice browning of clone 500–18 and 500–23 was distinctly stronger; however, their soluble solids content hardly differed from the control. In clone 500–2, all features of fruit quality were inferior to the original; nevertheless it had a good aspect in its compact growth habit (Table III).

The results of this fruit analysis indicate that mutations in the fruit characters of apple cultivar Fuji could have produced characteristics not only inferior, but also superior, to the original. Among seven mutants tested, mutants which could be directly used for practical cultivation would be clones 500–16 and 500–18. Probably 500–16 could be used as an attractive high quality fruit clone owing to its high content of soluble solids and its juicy texture. Although the high

TABLE II. MORPHOLOGICAL CHARACTERISTICS OF FRUIT OF SEVERAL INDUCED APPLE SPORTS

2 years average, 1974, 1975

IRB Clone	Fruit weight (g)	Fruit shape[a]	Lopsided[b]	Seed contents (No.)	No. of locules	Length of stalk (cm)
500–13	110.2	1.30	1.19	6.2	5.1	1.7
500–16	160.6	1.28	1.13	7.1	5.2	2.2
500–17	136.1	1.26	1.19	6.3	5.1	1.9
500–18	135.7	1.26	1.14	6.8	5.2	2.0
500–19	141.1	1.28	1.10	8.0	5.2	1.8
500–23	108.7	1.36	1.34	4.2	5.0	2.0
Fuji (cont.)	142.2	1.27	1.14	6.4	5.0	2.0
500–2	57.0	1.40	1.18	6.2	4.9	1.9
Ralls (cont.)	89.4	1.14	1.23	6.5	4.9	1.8

[a] Values of girth/height of fruit.
[b] Values of the longest height/ the shortest height of fruit.

incidence between spurriness of the tree and low content of soluble solids of the fruit was reported [17], the sweetness of fruits from clone 500–18 which has a spur-type growth habit was the same as the original. Consequently this clone migh be used as a blush-coloured fruiting spur type in practical cultivation. Both these clones are now under cultivation tests as V_3 propagations at the Fruit Tree Researc Station, Morioka. In the other five mutants, even though one character was superior to the original, other characters were undesirable, therefore their potenti value as parents in cross breeding should be tested further.

As Lapins [18] has already pointed out, most apple induced mutants produce fruit inferior to that of the non-mutated parent, and our mutants also showed this. Although their origin has not been presented here, a number of selected clones based on the change of vegetative organs have not differentiated flowering buds and have continued to produce only vegetative growth. Even though some of these drastic mutants have begun flowering and fruiting, the fruits were not worth of serious consideration because of inferior features of the fruit such as small,

TABLE III. FRUIT QUALITY OF SEVERAL INDUCED APPLE SPORTS
2 years average, 1974, 1975

IRB Clone	Firmness of texture (kg)	Acidity of juice (pH)	Sweetness of juice (Brix)	Degree of juice browning[a]
500–13	1.78	3.75	12.7	−
500–16	1.62	3.82	14.1	− −
500–17	2.05	3.80	13.4	+
500–18	1.17	3.74	13.2	++
500–19	1.93	3.81	13.2	+
500–23	2.17	3.83	13.6	++
Fuji	1.75	3.79	13.5	X
500–2	2.39	3.38	12.2	X
Ralls	1.60	3.29	12.5	X

[a] +: Dense; −: Pale; X: Non-varied.

misshapen fruit or bad fruit texture. In these experiments, two promising clones, 500–16 and 500–18, have arisen from short-term chronic irradiation treatment, not from the acute one. 500–13 has the same spur type as clone 500–18; however, its fruit quality was not acceptable. The cause might be attributed to the acute treatment. Consequently, for crops, including fruit trees, with the aim of mutation breeding in reproductive organs, chronic irradiation methods should be used.

3.   POLLEN VIABILITY

3.1. Materials and methods

Eleven apple clones originating from the cultivar Fuji, two clones from the cultivar Iwai (Summer Permains), and two clones from the cultivar Ralls, and their originals, were used as materials. Their growth habits, characteristics and their origin by mutation breeding at our Institute are presented in Table I.

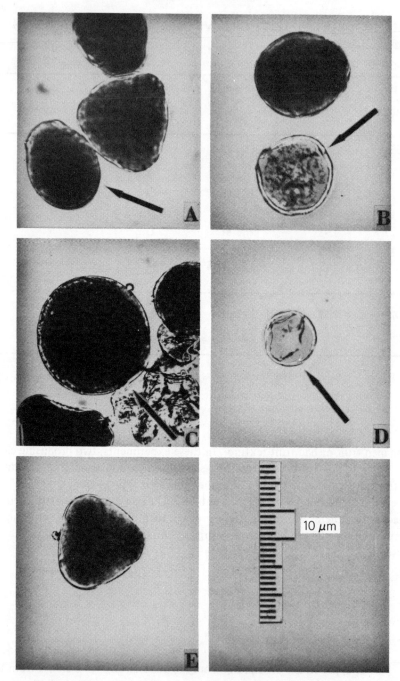

FIG.3. Pollen grain type of apple: A: A triangular and small grain (arrow), taken from 500−11; B: A round and small grain but not empty (arrow), from 500−12; C: A gigantic grain (arrow) from 500−13; D: A round, small and empty grain (arrow), from 500−24; E: The triangular normal grains taken from the cultivar Fuji.

At the bloom stage, 10 unopened flowers were collected from each tree (V_2), brought to the laboratory and stored in a Petri dish overnight. After anthesis, the pollen grains were stored at 5°C in a refrigerator. Microscopic observation of the pollen grains which were stained with aceto carmine was performed on their morphology and stainability. The ability of pollen germination of each clone was tested on agar medium containing 10% sucrose after overnight incubation at 22°C [19].

3.2. Results

Even in the control plants of varieties Fuji, Iwai and Ralls abnormal pollen grains which were round and small (Fig. 3B) were observed under the microscope among the normal grains which were triangular (Fig. 3E). In the induced clones, the abnormal morphology of the pollen grains was classified into three types: gigantic type (Fig. 3C), small type (Fig. 3B) and micro-type (Fig. 3D). Grains lacking cytoplasm and nuclei were always observed in micro-type grains. Within abnormal types, small grains lacking cytoplasm and nuclear material had the highest frequency; however, the occurrence of stainable small grains was quite frequent in 500–13 and 500–21. Gigantic and micro-type grains were occasionally produced in 500–12 and 500–13 although their percentage was not so high (Table IV).

Among clones from the cultivar Fuji, the most reduced in viability of pollen was 500–20, and it produced 84.7% abnormal grains over an average of three years (Table V). Clones 500–12, 500–21, 500–11 and 500–13 had also been induced by acute gamma-irradiation treatment. On the other hand, the spur-type clones, 500–18, 500–23, and dwarf-type clones, 500–17, 500–19, which had been induced chronically, seldom produced abnormal pollen grains, although their percentage was a little higher than that of the controls (Table V).

Among clones from the cultivar Iwai, a dwarf-type 500–0 and a spur-type 500–1 showed apparently lower fertility and resulted in a high production of abnormal pollen grains. In contrast, any marked difference in pollen fertility between induced clones from the cultivar Ralls and the original could not be confirmed although tree growth habits differed. No clones with complete male sterility could be detected; 500–14 was characterized as vegetative which did not differentiate any generative organs.

The results of observation over each of three years are shown in Table V. Although there was some annual variance in the production of abnormal pollen grains, the range of annual differences in fertility was not so great and was within about 10%. Therefore, the annual effect in the production of abnormal pollen grains was of little significance in the induced mutants of apple.

Most clones which produced highly viable pollen grains showed good germination on an agar bed. In contrast, the percentage of germinated grains and

TABLE IV. ABNORMAL TYPES OF POLLEN GRAINS AND THEIR FREQUENCY OF OCCURRENCE (1977)

IRB Clone	No. of pollen grains observed	Abnormal types					
		Stained grains			Empty grains		
		Gigantic (%)	Small (%)	Micro (%)	Normal (%)	Small (%)	Micro (%)
500−11	1300	0.2	8.4	0.0	0.0	4.1	0.0
500−12	906	0.2	7.1	0.0	0.1	52.5	0.3
500−13	1357	0.1	15.3	0.0	0.0	5.7	0.1
500−16	2180	0.1	0.3	0.0	0.0	4.4	0.0
500−17	1277	0.0	0.5	0.0	0.0	5.1	0.0
500−18	2800	0.1	1.1	0.0	0.1	2.6	0.1
500−19	967	0.0	0.1	0.0	0.0	3.6	0.0
500−21	509	0.0	37.1	0.0	0.0	7.1	0.0
500−23	650	0.0	1.4	0.0	0.2	4.5	0.2
500−24	1394	0.1	0.4	0.0	0.0	4.4	0.0
Fuji (cont.)	3477	0.0	0.4	0.0	0.0	4.1	0.0
500−1	238	0.0	5.0	0.0	0.0	58.4	0.0
Iwai (cont.)	554	0.0	1.3	0.0	0.0	5.8	0.0

the growth of pollen tube in abortive clones such as 500−1, 500−12 and 500−20 was low and weak, even though normally shaped pollen grains were obtained, but the population was low. Among the pollen types, the instances of germination of gigantic and micro-pollen grains could not be detected; however, stainable small grains with cytoplasm and nuclear material did germinate and their tubes could sufficiently penetrate into the agar (Fig. 4).

3.3. Discussion

It is well known that reduction in pollen viability arises from various causes such as the presence of a deficiency of chromosomes [20], formation of univalent-type chromosomes or some other type chromosomes at meiosis [21], defective

TABLE V. PERCENTAGE OF ABNORMAL POLLEN IN EACH OF THREE YEARS, 1975–77

Clone	1975		1976		1977		Average
			Mutants induced by acute irradiation				
500–13	1743[a]	14.3%	831	19.4%	1357	21.1%	18.4%
500–20	109	89.0	184	80.4	– – –[b]		84.7
500–21	219	32.9	382	26.7	284	44.2	34.6
500–11	– – –		– – –		1300	21.1	21.1
500–12	– – –		– – –		906	60.0	60.0
			Mutants induced by chronic irradiation				
500–16	704	9.3	634	3.4	2180	4.7	5.8
500–17	1871	7.6	864	5.8	1277	6.4	6.6
500–18	1832	7.9	877	4.4	1550	5.1	5.8
500–19	1297	5.5	1148	4.9	967	3.7	4.7
500–23	683	4.9	1092	10.7	650	6.2	7.3
500–24	– – –		332	9.3	1394	4.8	7.1
			Original				
Fuji	1677	5.0	1092	3.0	3477	4.5	4.2
500–0	464	43.0	376	74.3	554	63.5	60.2
500–1	712	54.1	576	37.7	– – –		45.9
Iwai (cont.)	681	2.9	734	6.9	238	7.0	5.6
500–2	386	6.0	586	11.3			8.7
500–3	– – –		137	0.7	– – –		0.7
Ralls (cont.)	361	2.2	137	11.7			7.0

[a] Number of pollen grains observed.
[b] Flowering not seen.

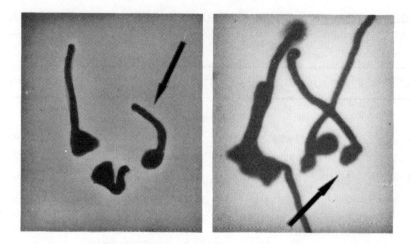

FIG.4. Germination of pollen grains of apple. The arrow represents tube growth of small round grains taken from IRB 500–13.

pollen wall formation, and the presence of sterile genes [22]. Mutations with partial gametic sterility have occurred more frequently than any other type of mutation in tree crops. In reviewing mutation breeding of tree crops, there are several reports on induced clones with reduced pollen viability [14, 20, 23]. From results of observation of V_1 propagates, Barritt and Eaton pointed out the relationship between the degree of pollen sterility in apple clones and irregular embryo sac development [24]. Also, they suggested the possibility of the selection of a clone with partially reduced fertility in order to reduce the labour of fruit thinning.

In the present study, using V_2 propagation, clones such as 500–11 and 500–12 showed a highly reduced pollen fertility and had not attained fruiting. Therefore, these clones with partial impotency would not have much value in saving labour in fruit thinning. On the other hand, 500–13 was characterized by 18.4% pollen fertility and would be a weak male sterile clone because of its good fruiting at out crossing. Its fruit characteristics have been presented elsewhere [12] Moreover, it has various type pollen grains such as gigantic and small. If abnormal grains take part in hybridization, unexpected segregation might be seen.

The relationship between the degree of pollen sterility expressed as a reduction in pollen stainability of each clone and its growth habit was negative. For instance, the range of abnormal grains production was from 60.2 to 5.8% among spur-type clones, and 84.7 to 0.7% from dwarf-type clones respectively.

Even a severe dwarf clone, 500–3, which has malformed and small variegated leaves, contains normally shaped grains although its production was less than that of the original cultivar Ralls.

It is of interest to note that the degree of pollen sterility of each clone correlated with its irradiation treatment at V_1. Data on abnormal pollen production indicates that a clone with severely reduced viability had originated from the acute gamma-ray treatment, not from the chronic (long-term exposure at a low dose rate). Moreover, these clones which showed a lower percentage of normal pollen than about 80% have been neither amply bearing nor shown favourable fruit quality, and would have no value for using in practical breeding programmes.

Chromosome aberration frequently affects the reduction capacity of an organism [25]. In Sugi, *Chryptomeria japonica*, all the six dwarf clones induced at IRB showed less than 50% pollen viability. However, chromosomal aberration at meiosis could not be detected [26]. Among induced clones of apple and cherries, abnormal behaviour of chromosome-like translocation and inverse hybridity at meiosis were reported and resulted in a higher production of round small pollen grains and less triangular shaped grains [20]. Although the cytological observations are not presented in this paper, several clones with reduced pollen viability and greater production of small round grains would have a similar chromosomal aberrant at meiosis.

4.   SECONDARY GROWTH AND LENGTH OF FIBRE CELLS

4.1. Materials and methods

Fourteen clones originating from the cultivar Fuji, two clones from the cultivar Iwai and two clones from the cultivar Ralls were used as material. Their origin and mutation breeding have been tabulated in a previous paper (Table I). Each clone consisted of potted five- or six-year-old trees on Maruba-Kaido rootstocks as V_2 propagations.

In middle of December 1976, the girth of the trunk at 5 cm above the bud union was measured by tape. Then 10 terminal one-year-old shoots were collected from each tree. Internodal length was determined on the middle section of each shoot. Microscopic observations were conducted on the cross-section of the middle and the basal portions of each shoot following Safranin staining. Also, the radial length of the secondary wood was measured with a micrometer and compared with the radial length of the bark. Samples for studying the elongation growth of the fibre cells were taken from the internodes of the middle portion of one-year-old shoots and the secondary wood was macerated with Jeffrey's solution after peeling the bark.

TABLE VI. CLONAL DIFFERENCES IN TREE GROWTH OF APPLE MUTANTS

IRB Clone	No. of trees	Tree height (cm)	Girth of trunk (cm)	Length of internode (cm)	Nature of branching
500−8	2[a]	214	13.1	1.0	More branching
500−9	2	164	9.2	1.4	Drooping
500−10	1	192	12.1	0.8	
500−11	2	153	13.2	1.0	
500−12	2	195	11.2	1.5	
500−13	2	174	12.3	2.3	Sparse
500−14	4	144	16.0	1.1	More branching
500−17	2	222	15.2	1.5	
500−18	2	189	13.0	2.0	Sparse
500−19	2	179	11.6	2.6	
500−20	2	204	14.8	1.2	More branching
500−21	2	250	15.5	2.1	Drooping
500−23	2	134	11.1	2.2	Sparse
500−24	2	174	10.4	2.3	Sparse
Fuji (cont.)	7	240	15.7	2.1	
500−0	2	165	17.2	1.9	
500−1	1	172	17.0	1.9	Sparse
Iwai (cont.)	4	147	14.7	1.7	
500−2	6	125	14.4	1.9	Sparse
500−3	3	214	12.9	1.4	
Ralls (cont.)	4	272	14.3	2.8	

[a] Potted five- or six-year-old trees.

4.2. Results

Selected clones having weak tree vigour can be classified either as spur type by reason of having many spur shoots and sparse branching, or dwarf type by reason of having weak thin shoots and narrow leaves. In the case of the cultivar Fuji, the internodal length of one-year-old shoots from spur-type clones such as 500–13, 500–18, 500–23 and 500–24 was not very different from that of the control. However, the internodal length of 500–2, a spur mutant from the cultivar Ralls, was apparently shorter than that of the control (Tables VI and VII).

Although the nature of branching was characterized by a low number of lateral shoots or sparsely in spur-type clones, clonal differences in dwarf-type clones were characterized by a wide variation. In particular, the branching of 500–14 trees was so severe that the length of each shoot was very short and resulted in losing apical dominance. Consequently, its tree form looked like a round crown if left unpruned. In clones 500–8, 500–9 and 500–20 the nature of branching was not the same as that of the control. Each shoot had a tendency to be thinner and drooping, resulting in appearance like a willow tree.

In Fig. 5, the relationship between the A-value which was found by comparing the length of current shoots with the basal diameter, and the B-value which was found by comparing the internodal length with the diameter on the middle portion of current shoots, is shown. It is evident that clones such as 500–10 and 500–14 having a higher A-value in comparison with the B-value have not as yet attained flowering up to 1977. On the other hand, clones such as 500–23 and 500–18 having a higher B-value in comparison with the A-value have attained the reproductive phase.

The range of wood formation on the middle portion of one-year-old shoots was very wide. In addition, without regard to spur type or dwarf type, all the selected clones except 500–19 had less wood formation than the control. Among them the lowest rate of wood formation was seen in 500–3, a dwarf mutant from the cultivar Ralls and 500–23, a spur-type mutant from the cultivar Fuji.

Clonal variation in the length of the fibre cells in the secondary wood was evident (Fig. 6). Also, almost all clones except 500–13 have apparently shorter fibre cells than the control. Several mutants from the cultivar Fuji, such as dwarf-type clones 500–8, 500–9 and 500–14, had much shorter cells compared with those of the control and almost all the fibre cells had a tracheal nature in differentiation. On the contrary, 500–20, a dwarf-type clone, had the same or a little longer fibre cells compared with those of the control. Marked differences in length of fibre cells were not evident among clones of spur-types from the cultivar Fuji such as 500–13, 500–18 and 500–23. However, a spur-type clone, 500–2, originating from the cultivar Ralls, had shorter fibre cells in comparison with the control (Fig. 6).

TABLE VII. CLONAL DIFFERENCES IN THE FORMATION OF WOOD
IN CURRENT SHOOTS

Clone	Half diameter of middle part of shoot (A)[a] (mm)	Half diameter of pith (mm)	Radial width of wood (B) (mm)	Radial width of bark (mm)	Rate of wood formation (B/A × 100) (%)
500−8	1.58	0.57	0.36	0.65	23.6
500−9	1.45	0.54	0.37	0.53	25.8
500−10	1.69	0.56	0.53	0.59	31.6
500−11	1.69	0.62	0.47	0.60	27.7
500−12	1.82	0.71	0.47	0.65	25.7
500−13	1.96	0.73	0.57	0.67	28.8
500−14	1.82	0.67	0.49	0.66	27.0
500−17	1.67	0.60	0.50	0.57	30.0
500−18	1.88	0.73	0.50	0.65	26.6
500−19	1.81	0.63	0.65	0.54	35.6
500−20	1.78	0.58	0.48	0.72	27.0
500−21	1.67	0.58	0.50	0.57	29.7
500−23	1.93	0.70	0.52	0.72	19.3
500−24	2.29	0.79	0.67	0.83	29.2
Fuji	1.70	0.62	0.54	0.54	32.4
500−0	1.84	0.69	0.48	0.68	25.7
500−1	1.74	0.62	0.52	0.60	30.0
Iwai	1.41	0.38	0.47	0.56	33.0
500−2	2.15	0.60	0.60	0.89	27.7
500−3	1.48	0.33	0.31	0.60	20.6
Ralls	1.97	0.67	0.69	0.62	34.8

[a] Data were taken from 10 representative terminal shoots.

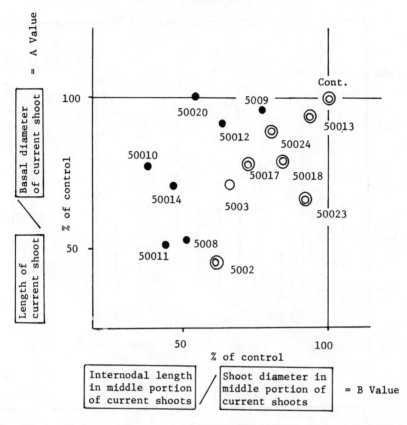

FIG.5. *Relationship between the nature of shoot growth and fruiting ability in induced clones of apple. Double circles represent a clone which had attained fruiting in 1977. Single circle represents a clone which had attained flowering but had not yet fruited. Closed circles represent a clone which had not attained the reproductive phase.*

4.3. Discussion

It has been pointed out that ionizing radiation treatment in apple has a relatively strong effect on induction of mutants having reduced tree size [8, 9, 27–30]. However, no typical examples of desirable tree forms for fruit cropping and dense planting were obtained. Lapins recommended using those mutants characterized by reduced size, shortened internodal lengths in shoots, thick shoots, and reduced number of vegetative shoots by an increased number of fruiting spurs [18]. In the present experiment, mutants having reduced tree size could be basically classified as spur types because of sparse branching, more spur shoots, plump shoots, and dwarf types because of slender shoots and small leaf area. Moreover, these characteristics can be genetically independent. For example, 500–2, a spur-type

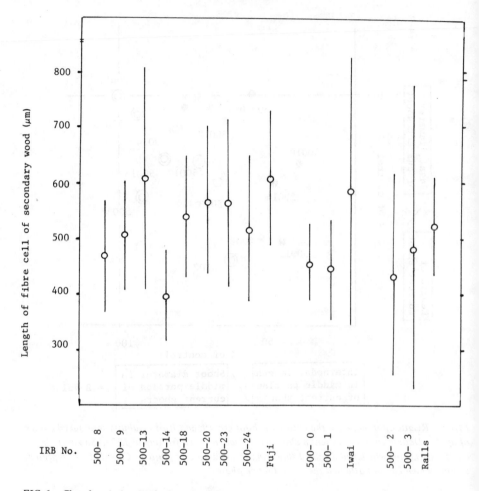

FIG.6. *Clonal variation in the length of fibre cells in the wood. Data taken from 300 cells in one-year-old shoots. Each circle represents the average length of a clone and the bar represents its standard deviation.*

clone from the cultivar Ralls has short internodes and short shoots which is characteristic of the dwarf-type in many instances. However, there was no difference in internodal length between 500—18, a spur-type clone, and the original Fuji. Consequently, in the case of the former clone, a dwarf mutation can be added to a mutation in spurriness of the tree.

Out of 13 clones from the cultivar Fuji, two clones from the cultivar Iwai and two clones from the cultivar Ralls nine clones have attained flowering and fruiting as of 1977. In the remaining eight clones, no floral organ formation was detected.

From analysis of the elongation growth and the increment growth in one-year-old shoots, these vegetative-type clones were characterized as acropetally weak in increment growth which resulted in slender shoots. It is pointed out that these clones have no value for fruit improvement except for possible use as rootstocks although their vegetative reproductivity is still obscure.

The rate of secondary wood formation in one-year-old shoots was low in all the clones in comparison with the controls except for one clone. Moreover, clonal variation in the length of the fibre cells was recognizable. Results of previous studies on increment growth suggest that cambial activity, including secondary xylem development, is regulated by the auxin and gibberellin systems controlled by meristems [31, 32]. Moreover, a more direct line of evidence implicating gibberellins and abscisic acid with the regulation of apple shoot extension comes from a recent study in which these substances have been exogenously applied to apple trees [33]. Following these theories on hormonal regulation concerning increment growth and shoot extension growth, lower levels of auxin and gibberellins should be detectable in the cambium of selected mutants by reason of the fact that they showed weak elongation growth and weak increment growth of the wood. Also, it could be postulated that the shorter cell length of the wood fibres following elongation growth of the cambial fusiform initials is correlated with a lower concentration of gibberellins and auxin in the cambium [34]. In the present work, it is proposed that a clone like 500–14, which has a much lower rate of wood formation and shorter fibre cell length, could be a deficient mutant in the production of gibberellins and auxin.

5.   HISTOLOGY OF SHOOT APICES

5.1. Materials and methods

In order to determine the ploidal levels in the apical meristem, median longitudinal sections of the shoot apex were prepared by the ordinary paraffin methods and stained with Delifield's haematoxylin. Ten shoots of each mutant were used. Within them, the three representative figures in which the fine zonal structure can be ascertained were selected. For 10 cells composing each layer of the apical dome, the diameter length of the nucleus was measured with a micrometer.

Current growth shoots were collected from potted trees during July 1977 and were stored in Formarin acetic acid alcohol solution until sectioning. The plant materials used in this experiment were 15 mutant clones originating from the cultivar Fuji, one mutant clone from the cultivar Iwai and two mutant clones from the cultivar Ralls and their originals.

TABLE VIII. CLONAL COMPARISON OF CELL NUCLEAR VOLUME IN THE APICAL MERISTEM AMONG INDUCED MUTANTS OF APPLE

IRB Clone	Layers in apical meristem[a]					Average (μm)
	L−1 (μm)	L−2 (μm)	L−3 (μm)	L−4 (μm)	L−5 (μm)	
500−8	3.3 ± 0.1	3.3 ± 0.1	3.3 ± 0.1	3.3 ± 0.0	3.3 ± 0.1	3.3 ± 0.1
500−9	3.9 ± 0.1	4.1 ± 0.1	3.9 ± 0.1	3.9 ± 0.1	4.0 ± 0.1	4.0 ± 0.1
500−10	3.9 ± 0.1	4.2 ± 0.1	3.9 ± 0.1	3.8 ± 0.1	4.1 ± 0.1	4.0 ± 0.1
500−12	2.8 ± 0.1	2.8 ± 0.1	3.0 ± 0.1	3.2 ± 0.2	2.6 ± 0.4	2.9 ± 0.2
500−13	4.3 ± 0.0	4.6 ± 0.0	4.2 ± 0.0	4.0 ± 0.0	4.0 ± 0.0	4.2 ± 0.1
500−14	4.0 ± 0.1	3.9 ± 0.1	3.8 ± 0.1	3.8 ± 0.1	3.6 ± 0.1	3.8 ± 0.1
500−15	4.0 ± 0.1	3.9 ± 0.1	3.7 ± 0.3	3.7 ± 0.1	3.7 ± 0.1	3.8 ± 0.1
500−16	3.9 ± 0.1	3.8 ± 0.1	3.5 ± 0.1	3.6 ± 0.1	3.6 ± 0.1	3.7 ± 0.1
500−17	4.0 ± 0.1	3.9 ± 0.1	3.8 ± 0.1	3.7 ± 0.1	3.8 ± 0.1	3.8 ± 0.1
500−18	4.2 ± 0.1	3.9 ± 0.2	4.0 ± 0.1	4.0 ± 0.3	3.8 ± 0.2	4.0 ± 0.2
500−19	3.7 ± 0.3	3.6 ± 0.3	3.9 ± 0.4	3.7 ± 0.3	4.1 ± 0.4	3.8 ± 0.3
500−20	3.5 ± 0.1	3.5 ± 0.4	3.6 ± 0.3	3.5 ± 0.1	3.7 ± 0.9	3.6 ± 0.4
500−21	3.7 ± 0.0	3.8 ± 0.1	3.8 ± 0.1	3.8 ± 0.0	3.8 ± 0.0	3.8 ± 0.0
500−23	3.7 ± 0.0	3.6 ± 0.1	3.6 ± 0.0	3.7 ± 0.1	3.7 ± 0.1	3.7 ± 0.1
500−24	3.8 ± 0.1	3.9 ± 0.1	3.6 ± 0.1	3.7 ± 0.1	3.7 ± 0.1	3.7 ± 0.1
Fuji	4.0 ± 0.2	4.3 ± 0.1	4.3 ± 0.2	4.0 ± 0.1	3.9 ± 0.1	4.1 ± 0.1
500−1	3.8 ± 0.1	3.8 ± 0.1	3.7 ± 0.1	3.8 ± 0.0	3.7 ± 0.0	3.8 ± 0.1
Iwai	4.0 ± 0.1	4.2 ± 0.1	4.0 ± 0.1	3.7 ± 0.1	3.9 ± 0.1	4.0 ± 0.1
500−2	4.3 ± 0.0	4.4 ± 0.1	4.1 ± 0.0	4.1 ± 0.0	4.1 ± 0.0	4.2 ± 0.0
500−3	4.0 ± 0.1	4.1 ± 0.1	4.1 ± 0.1	3.9 ± 0.0	4.0 ± 0.0	4.0 ± 0.1
Ralls	3.9 ± 0.0	4.5 ± 0.4	4.4 ± 0.1	4.2 ± 0.1	4.2 ± 0.1	4.2 ± 0.0

[a] Diameter length of the cell nucleus.

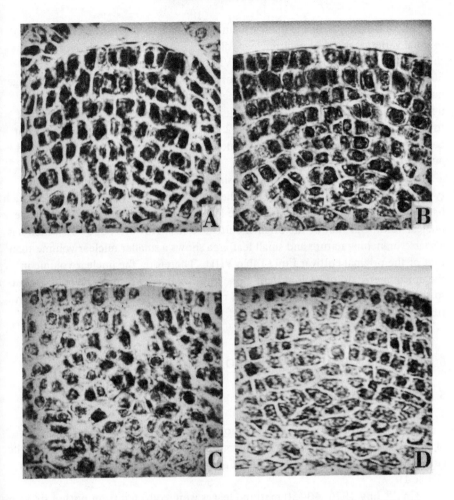

*FIG.7. Shoot apices of apple mutants and their original cultivar Fuji. A: 500–14;
B: 500–9; C: 500–18; D: Control.*

5.2. Results and discussion

In Table VIII, all the mutant clones with weak tree vigour show a smaller
volume of the nucleus in the apical meristem than their original cultivars except
for mutant clones 500–2, 500–13 and 500–18. The smallest nuclear volume
was detected in the apical dome of 500–12 which is a severe dwarf-type mutant.
Also, the largest nuclear volume was detected in the apical dome of 500–13 which
is a spur-type mutant. It seems therefore that spur-type mutant clones have the
same nuclear volumes as the originals or perhaps a little larger.

In comparison with the nuclear volumes of each histogenic layer, all the original varieties show a larger nuclear volume in the subepidermal layer, L-2, than in the epidermal layer, L-1 (Table VIII, Fig. 7). On the other hand, clone 500—18, which is a preferable mutant for fruit cropping [35], shows a much larger nuclear volume in L-1 cells than in L-2 cells (Table VIII, Fig. 7). Although the mitotic division figures in L-1 cells of this clone could not be observed, it seems to be a polyploidal chimera such as 4-2-2. If this is true, its spurriness might not be transmitted in a cross breeding test such as Lapins recognized on the induced mutant of McIntosh [36].

Generally speaking, a clone which shows sparse branching, such as a spur type, has a very large spherical area of the shoot apex. For instance, 500—2 which has a sparse branching nature and a large leaf area was characterized by the same nuclear volume as that of the original cultivar Ralls. However, 500—15 which has a sparse branching nature and small leaf area shows a smaller nuclear volume than that of the original cultivar Fuji (Table VIII). Therefore, the nuclear volume in the apical meristem of each mutant clone with weak growth habits would correspor to the relative volume of the organs and tissues such as the leaf, fruit, corolla and fibre cell length in wood, and so on [37].

6.   COLOUR REACTION WITH IODOACETIC ACID IN GREEN LEAVES

6.1. Materials and methods

Two clones of the cultivar Iwai, two clones of the cultivar Ralls, 11 clones of V_2 propagations of the cultivar Fuji and three clones of V_3 propagations of the cultivar Fuji were used as plant material and were compared with their originals (Table I).

On 28 July 1976, 40—50 matured leaves were collected from potted six-year-old trees grafted on Maruba-Kaido rootstock. Leaf samples were collected between 10:00 to 12:00 and were brought immediately to the laboratory and stored in a refrigerator at 0°C until the leaf discs were punched. Twenty discs, 1 cm in diameter, were punched from each leaf, and were floated on an aqueous solution of mono-iodoacetic acid in a concentration of 10^{-4} M. All discs were placed abaxial side up. The covered Petri dishes, 8.5 cm in diameter, having 20 discs floating on a 20 ml solution, were stored for four days (exactly 96 ± 0.5 hours) in an incubator at 27°C. The solution was then filtered.

It is recognized that there is a high correlation between the degree of browning of green leaves and the amount of the brown coloured substance developed during incubation in solution [38]. Comparisons between clones were conducted only on measurement of the solution so far as this experiment is concerned. The intensity of the red colouring in the solution was based on a reading of the aL

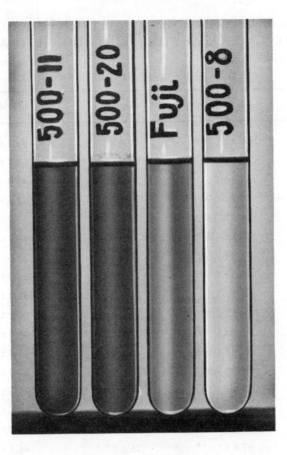

FIG.8.   Colouring solution developed from green leaf discs with iodoacetic acid and a clonal comparison with the original cultivar Fuji. In 500—11 and 500—20, the solution is darker than that of the original, and in 500—8 it is brighter.

value with a colour and colour difference meter and the intensity of the yellow colour was based on the bL value. The degree of the brightness in the solution was determined with the reading of the L value on the meter. Ten replications were conducted on each tree and analysed statistically.

6.2. Results and discussion

Iodoacetic acid is nonspecific in its action; however, it can inhibit several enzymes in the Emben-Meyerhof-Parnas pathway [39]. This reagent is able to

TABLE IX. CLONAL COMPARISON ON THE INTENSITY OF COLOUR
DEVELOPED IN GREEN LEAF WITH IODOACETIC ACID IN APPLE
CULTIVAR FUJI[a]

Clone tree	L-value[b]	aL-value	bL-value
500–8–3	65.4 ± 1.1**	20.4 ± 0.7*	44.7 ± 0.8**
500–8–4	64.4 ± 1.7*	20.3 ± 0.8**	44.5 ± 1.2*
500–9–1	63.9 ± 1.6	26.4 ± 0.4	43.9 ± 1.1
900–9–2	61.5 ± 1.7	21.4 ± 0.8	42.1 ± 1.2
500–11–1	46.5 ± 2.5**	26.7 ± 0.8*	31.7 ± 1.7**
500–11–2	42.8 ± 1.5**	25.0 ± 0.4	29.3 ± 1.0**
500–12–1	52.8 ± 1.0**	22.5 ± 0.6	36.5 ± 0.7**
500–121	55.6 ± 1.7	23.4 ± 0.4	38.3 ± 1.1
500–131–4	56.0 ± 1.4	22.3 ± 1.6	38.3 ± 0.9
500–131–5	52.3 ± 2.1*	25.4 ± 0.8	36.1 ± 1.4*
500–14–1	53.6 ± 1.1**	27.2 ± 0.8**	36.7 ± 1.3**
500–141–4	49.0 ± 1.8**	25.0 ± 0.4	34.4 ± 1.3**
500–16–1	54.8 ± 1.3	24.6 ± 0.9	37.8 ± 0.9
500–18–1	57.9 ± 1.1	26.8 ± 0.7**	39.6 ± 0.8
500–18–2	56.9 ± 0.9	24.4 ± 0.4	39.5 ± 0.7
500–19–1	61.0 ± 2.3	24.6 ± 1.3	41.7 ± 1.6
500–21–1	67.5 ± 1.0**	26.1 ± 1.1	46.3 ± 0.7**
500–23–2	61.8 ± 1.5	25.9 ± 0.6	42.3 ± 1.1
500–24–1	46.2 ± 1.7**	24.7 ± 0.5	31.7 ± 1.2**
500–24–2	46.8 ± 2.0**	26.0 ± 1.1	31.9 ± 1.4**
Fuji (cont.)	59.6 ± 1.2	23.9 ± 0.5	40.9 ± 1.4

[a] Data from the colour and colour difference meter (Nihon Denshoku Kogyo Co.). The
L-value represents brightness of solution and the higher the numbers, the brighter the
solution. The aL and bL-values represent the degree of red and yellow colouring. The
higher the numbers, the greater the intensity.

[b] *, ** indicate the significant variance ratios at $P < 0.05$ and $P < 0.01$.

turn green apple leaves to a brown colour and the brown coloured substrate exuded from the leaf tissue during incubation if they are immersed in its aqueous solution [38]. From the results of a comparison between the cultivar Iwai or the cultivar Ralls and their mutants, it was readily recognizable that the amount of the brown coloured substance developed in green leaf discs treated with iodoacetic acid and dissolved into solution during incubation differs (Fig. 8). Moreover, the degree of colouring in the solution of dwarf-type mutants was redder and less yellow compared with compact spur-type mutants in every case.

On the comparison of mutants derived from the cultivar Fuji, the intensity of the colour of the solution was almost always darker than that of the original. The clones which showed a statistically signficant difference in the bL values and the L value measurements from the original were 500–8, 500–11, 500–12, 500–14, 500–21 and 500–24 (Table IX). Among these, 500–8 and 500–21 showed a much brighter colour than the original (Fig. 8).

It is interesting to note that the tree vigour of these clones which had been selected as severe dwarf-type clones at the V_1 stage was very weak (Table IX). In spur-type clones such as 500–18, 500–23 and 500–24, the intensity of the red colour in solution was darker in comparison with the original; however, the intensity of the yellow colour in solution was brighter in 500–23 and darker in 500–18 and 500–24 (Table IX).

Although it was reported that iodoacetic acid could enhance the anthocyanin development in apple skins at low concentrations [39], and act as an inhibitor for anthocyanin development in Saxifolia leaves at high concentrations [40], the brown coloured substance in apple leaf was completely precipitated with lead acetate at a concentration of 10^{-2} M if the solution was mixed with it in proportions of the same volume. The solution also removed the colour if the pH was adjusted to a high acid range with hydrochloric acid. Therefore, it is assumed that the colouring matter might be a polyphenol compound [41]. Siegelman had detected the existence of 1-epicatechin in apple fruit tissue [42]; however, the chemical nature of the precipitated compound in the present experiment is obscure.

It is premature to determine whether the clonal differences in the degree of colouring with iodoacetic acid could be attributed to the detection of any phenotypic variation in the apple. However, the experimental results lead us to believe there is a relationship between the varying production of brown coloured substances, polyphenol compounds which might act as a growth inhibitor [43, 44], and dwarfism. Thus, the colour reaction of green leaves with iodoacetic acid might be used as a method for the early detection of mutants by a chemical procedure. From our results, it is not clear whether clonal differences in the development of the brown coloured substance were caused by clonal differences in the enzymatic activity of the green leaf discs or by clonal differences in the amount of the substrate. Polyphenol oxidase activity will be measured with each apple mutant in future experiments.

REFERENCES

[1] SHAMEL, A.D., POMEROY, C.S., Bud mutations in horticultural plants., J. Hered. **27** (1936) 487.

[2] EINSET, J., Spontaneous polyploidy in cultivated apples, Proc. Am. Soc. Hort. Sci. **59** (1952) 291.

[3] EINSET, J., PRATT, C., Spontaneous and induced apple sports with misshapen fruit, Proc. Am. Soc. Hort. Sci. **73** (1959) 1.

[4] GRANHALL, I., GUSTAFSSON, Å., NILSSON, F., OLDEN, E.J., X-ray effects in fruit trees, Hereditas **35** (1949) 269.

[5] GRANHALL, I., X-ray mutations in apple and pears, Hereditas **30** (1953) 149.

[6] BISHOP, C.J., Radiation-induced fruit color mutations in apples, Can. J. Genet. Cytol. **1** (1959) 118.

[7] BISHOP, C.J., Mutations in apples induced by X-radiation, J. Hered. **45** (1954) 99.

[8] LAPINS, K.O., Compact mutants of apple induced by ionizing radiation, Can. J. Plant Sci. **45** (1965) 117.

[9] NISHIDA, T., The induced semi-dwarf mutants of apple by gamma irradiation, Technical News No.2 (1969) 1.

[10] PRATT, C., WAY, R.D., OURECKY, D.K., Irradiation of color sports of Delicious and Rome apples, J. Am. Soc. Hort. Sci. **97** (1972) 268.

[11] ZAGAJA, S.W., PRZYBYŁA, A., "Gamma-ray mutants in apples", Induced Mutations in Vegetatively Propagated Plants (Proc. Panel Vienna, 1972), IAEA, Vienna (1973) 35.

[12] IKEDA, F., Radiation induced fruit color mutation in apple var. Fuji, Technical News No. 15 (1974) 1.

[13] IKEDA, F., Induced mutation in dwarf growth habits of apple trees by gamma rays and its evaluation in practical uses, Gamma Field Symp. **16** (1977) 63.

[14] LAPINS, K.O., Mutation of Golden Delicious apple induced by ionizing radiation, Can. J. Plant Sci. **51** (1971) 123.

[15] NISHIDA, T., Induction of the somatic mutations in deciduous fruit trees by gamma irradiation, Gamma Field Symp. **12** (1973) 1.

[16] SADAMORI, S., et al., New apple variety "Fuji", Bull. Hort. Res. Sta. M.A.F., Series C. No. 1 (1963) 1.

[17] MEHERIUK, M., FISHER, D.V., LAPINS, K.O., Some morphological and physiological features of several red Delicious apple sports, Can. J. Plant Sci. **53** (1973) 335.

[18] LAPINS, K.O., "Induced mutations in fruit trees", Induced Mutations in Vegetatively Propagated Plants (Proc. Panel Vienna, 1972), IAEA, Vienna (1973) 1.

[19] SUSA, T., Studies on self- and cross-pollination in apple, J. Hort. Assoc. Jpn **5** (1934) 194 (in Japanese).

[20] WHELAN, E.D.P., HORNBY, C.A., Meiotic abnormalities and pollen viability in *Prunus avium* L. cv. Lambert, J. Am. Soc. Hort. Sci. **94** (1969) 263.

[21] WHELAN, E.D.P., HORNBY, C.A., LAPINS, K.O., Variation in pollen viability and induced chromosomal aberrations among and within propagates of irradiated *Prunus avium* L. cv. Lambert, J. Am. Soc. Hort. Sci. **95** (1970) 763.

[22] HEILBORN, O., Reduction division, pollen lethality and polyploidy in apple, Acta Hort. Berg. **11** (1935) 129.

[23] OHBA, K., MAETA, T., Induction of somatic mutations and cross fertility of the mutants in Sugi, *Cryptomeria japonica* D. Don., Gamma Field Symp. **12** (1973) 19.

[24] BARRITT, B.H., EATON, G.W., Embryo sac development in relation to pollen fertility and seed set in apple clones produced from ^{60}Co irradiated dormant buds, Radiat. Bot. **6** (1966) 589.

[25] SPARROW, A.H., "Types of irradiation and their cytogenetic effects", Mutation and Plant Breeding, NAS/NRC 891 (1961) 55.

[26] DERMEN, H., BAIN, H.F., A genetical cytohistological study of cholchicine polyploidy in Cranberry, Am. J. Bot. 31 (1944) 451.

[27] CAMPBELL, A.I., LACEY, N.D., Compact mutants of Bramleys Seedling apple induced by gamma radiation, J. Hort. Sci. 48 (1973) 397.

[28] LAPINS, K.O., Tree growth habits in radiation induced mutants of McIntosh apple, Can. J. Plant Sci. 49 (1969) 483.

[29] POLL, L., The production and growth of young apple compact mutants induced by ionizing radiation, Euphytica 23 (1974) 521.

[30] VISSER, T.J., VERHAEGH, J., DEVRIES, D.P., Preselection of compact mutants induced by X-ray treatment in apple and pear, Euphytica 20 (1971) 195.

[31] LARSON, P., "Auxin gradients and the regulation of cambial activity", Tree Growth (KOZLOWSKI, T.T., Ed.), Ronald Press (1962) 97.

[32] WAREING, P.F., HANNEEY, C.E.A., DIGHY, J., "The role of endogenous hormones in cambial activity and xylem differentiation", The Formation of Wood in Forest Trees (ZIMMERMANN, M.H., Ed.), Academic Press (1964) 323.

[33] ROBITAILLE, H.A., CARSON, R.F., Gibberellic and abscisic acid-like substances and regulation of apple shoot extension, J. Am. Soc. Hort. Sci. 101 (1976) 388.

[34] KATO, J., "Physiological action of gibberellins", Gibberellins (TAMURA, S., Ed.), Tokyo Univ. Press (1969) 282.

[35] IKEDA, F., NISHIDA, T., Comparisons of fruit characteristics of several apple sports with their originals, J. Jpn Soc. Hort. Sci. 45 Suppl.1 (1977) 8.

[36] LAPINS, K.O., Segregation of compact growth types in certain apple seedling progenies, Can. J. Plant Sci. 49 (1969) 765.

[37] KURAISHI, S., Hormonal aspect of dwarfism, Gamma Field Symp. 15 (1976) 33.

[38] IKEDA, F., Studies on the early detection technique for useful mutation. 1. Reaction to iodoacetic acid in apple green leaf, Annual Rep. (1975) 63 (in Japanese).

[39] FAUST, M., Physiology of anthocyanin development in McIntosh apple. 1. Participation of pentose phosphate pathway in anthocyanin development, Am. Soc. Hort. Sci. Proc. 87 (1965) 1.

[40] EBERHARDT, F., Über die Beziehungen zwischen Atmung und Anthocyansynthese, Planta 43 (1954) 253.

[41] HARBOREN, J.B., Comparative Biochemistry of the Flavonoides, Academic Press (1967).

[42] SIEGELMAN, H.W., Detection and identification of polyphenol oxidase substrates in apple and pear skins, Arch. Biochem. Biophys. 56 (1955) 97.

[43] FURUYA, M., GALSTON, A., STOWE, B., Isolation from peas of co-factors and inhibitors of indolyl-3-acetic acid oxidase, Nature (London) 193 (1962) 456.

[44] MARTIN, G.C., STAHLY, E.A., Endogenous growth regulating factors in bark of EM IX and XVI apple trees, Am. Soc. Hort. Sci. Proc. 91 (1967) 31.

PROGRESS IN SELECTION FOR SODIUM CHLORIDE, 2,4-D DICHLOROPHENOXY ACETIC ACID (2,4-D) AND STREPTOMYCIN TOLERANCE IN *Citrus sinensis* OVULAR CALLUS LINES*

J. KOCHBA, P. SPIEGEL-ROY
Volcani Centre, ARO,
Institute of Horticulture,
Division of Fruit Breeding
    and Genetics,
Bet-Dagan,
Israel

Abstract

PROGRESS IN SELECTION FOR SODIUM CHLORIDE, 2, 4-D DICHLOROPHENOXY ACETIC ACID (2, 4-D) AND STREPTOMYCIN TOLERANCE IN *Citrus sinensis* OVULAR CALLUS LINES.

Citrus sinensis (cultivar Shamouti) nucellar embryogenic callus lines with greatly increased tolerance to salinity (NaCl), 2, 4-D and streptomycin were selected. Selected lines were found stable after removal of selection pressure. Gamma irradiation at $8-16$ kR was also employed and found to speed up selections. Embryos from NaCl and 2, 4-D tolerant lines also showed increased tolerance. Embryogenesis in selected lines, suppressed during selection procedures, was regained by growing cultures in the presence of galactose or lactose as the sole carbon source. A schedule was worked out furthering development of embryos into plantlets. Conditions for adventive shoot formation from embryonic shoot segments were established, thus allowing cloning of embryos. A procedure was worked out for suspension culture and agar plating of cell groups.

INTRODUCTION

The application of tissue and cell cultures as an aid in breeding *Citrus* is justified because of limitations to hybridization imposed by apomixis (nucellar embryony) on the one hand and avoidance of chimera formation, frequently encountered in mutation breeding, on the other hand [1, 2]. The advantages of cell cultures for selection of mutants have been extensively reviewed [3–6], and progress and problems in this field were recently critically evaluated [7].

---

* Research supported by the IAEA under Research Contract No. 1273.

At the last research co-ordination meeting sponsored by the IAEA, in Skierniewice, Poland (1978), we reported [8] on the selection of salinity (NaCl) and 2, 4-D tolerant lines from Shamouti orange *(Citrus sinensis)* nucellar embryogenic callus. Lines obtained from non-irradiated callus were described and the beginning of selection from gamma-irradiated tissue was commented on. Studies were continued, and new NaCl and 2, 4-D tolerant lines from gamma-irradiated callus were selected and tested. In addition, streptomycin-tolerant lines have been selected from non-irradiated callus.

Much attention was given to the regeneration of embryos and establishment of plants from selected lines. Although intact plants were not yet tested, elevated tolerance of embryos from selected lines was found, and a satisfactory schedule for plant regeneration and cloning was worked out.

Considerable progress has been made during the period of this co-ordinated research programme. The conditions for embryogenesis in *Citrus callus* were established [8–14], and findings were then applied in the mutant selection programme. The work, however, is far from being completed. The crucial phase has now been reached in which we shall test whether traits in cell cultures are manifested in intact plants. Selections aiming to recover disease resistance in important *Citrus* rootstock types will be attempted, and if successful will contribute greatly to the establishment of disease-free clonal rootstocks.

MATERIALS AND METHODS

The selection studies were carried out with ovular callus of *Citrus sinensis*, cultivar Shamouti. Culture conditions have been described earlier [9] and the selection protocol was detailed at the meeting in Skierniewice in 1978. For irradiation, 8, 12 and 16 kR were applied at a dose rate of 80 kR/h, from a caesium gamma source at the Nahal Soreq Nuclear Research Centre, Israel. The selective concentration was either 5 g/l NaCl, 1×10^{-5} M 2, 4-D or 175 mg/l streptomycin sulphate respectively. For final tests of tolerance elevated concentrations of the selecting agent were applied, either 10 g/l NaCl, 1×10^{-4} M 2, 4-D or 1000 mg/l streptomycin sulphate respectively.

RESULTS AND DISCUSSION

Callus lines tolerant to salinity

Four lines were selected from gamma-irradiated calli and one line from non-irradiated controls, after recurrent selection over 10 culture passages each of five weeks. All selected lines showed significantly better growth on 5 g/l NaCl. Three of them yielded also significantly better growth on 10 g/l NaCl (Table I).

TABLE I. PERFORMANCE OF SELECTED SHAMOUTI ORANGE CALLUS LINES, OBTAINED AFTER GAMMA IRRADIATION, ON INCREASING NaCl CONCENTRATIONS

Callus line	NaCl concentration		
	0	5 g/l	10 g/l
L-1 control	725 ± 57^a	430 ± 80	135 ± 34
R-3 (0 kR)	800 ± 80	800 ± 133	295 ± 36
R-2 (8 kR)	665 ± 42	525 ± 106	235 ± 20
R-5 (12 kR)	720 ± 78	700 ± 121	335 ± 78
R10 (12 kR)	770 ± 105	765 ± 139	210 ± 64
R-14 (16 kR)	660 ± 68	845 ± 140	195 ± 35

[a] mg fresh weight/culture.

TABLE II. TEST OF STABILITY OF TOLERANCE TO NaCl FOLLOWING ABSENCE OR CONTINUOUS PRESENCE OF SELECTION PRESSURE

Subculture	Growth (mg) of lines on			
	5 g/l NaCl		10 g/l NaCl	
	Continuous	Discontinuous	Continuous	Discontinuous
	R-10 (12 kR)			
1st	420 ± 30^a	380 ± 40	110 ± 47	140 ± 30
2nd	640 ± 151	365 ± 31	200 ± 17	255 ± 21
3rd	410 ± 55	360 ± 42	210 ± 64	230 ± 17
	R-14 (16 kR)			
1st	560 ± 72	550 ± 80	120 ± 21	135 ± 15
2nd	820 ± 113	660 ± 70	165 ± 18	175 ± 14
3rd	570 ± 70	560 ± 26	195 ± 35	135 ± 39

[a] mg fresh weight of culture.

TABLE III. PERFORMANCE OF SELECTED SHAMOUTI ORANGE CALLUS LINES, AFTER GAMMA-IRRADIATION, ON INCREASING 2, 4-D CONCENTRATIONS

Callus line	2, 4-D concentration (molar)						
	0	1×10^{-5}	2×10^{-5}	3×10^{-5}	4×10^{-5}	5×10^{-5}	1×10^{-4}
Control	932 ± 95	316 ± 113	215 ± 31	140 ± 31	135 ± 4	128 ± 2	110 ± 6
R1 (0 kR)	1022 ± 78[a]	837 ± 123	690 ± 61	535 ± 66	515 ± 87	432 ± 53	425 ± 56
R1 (8 kR)	1185 ± 31	917 ± 47	900 ± 48	760 ± 96	670 ± 52	602 ± 68	520 ± 86
R10 (12 kR)	1102 ± 101	670 ± 116	750 ± 32	700 ± 30	545 ± 42	412 ± 33	110 ± 13
R 5 (16 kR)	945 ± 113	862 ± 95	490 ± 55	165 ± 13	158 ± 2	182 ± 21	100 ± 13

[a] mg fresh weight/culture.

TABLE IV. TEST OF STABILITY OF TOLERANCE TO 2, 4-D FOLLOWING ABSENCE OR CONTINUOUS PRESENCE OF SELECTION PRESSURE

| | Growth (mg) of lines on | | | |
| | 1×10^{-5} 2, 4-D | | 5×10^{-5} 2, 4-D | |
Subculture	Continuous	Discontinuous	Continuous	Discontinuous
		R-1 (8 kR)		
1st	760 ± 61[a]	790 ± 52	440 ± 60	515 ± 57
2nd	725 ± 41	765 ± 50	390 ± 60	545 ± 48
3rd	655 ± 38	545 ± 25	305 ± 45	475 ± 31
		R-10 (12 kR)		
1st	885 ± 128	620 ± 46	410 ± 63	145 ± 18
2nd	585 ± 85	595 ± 38	230 ± 45	260 ± 55
3rd	415 ± 30	445 ± 34	230 ± 41	240 ± 23

[a] mg fresh weight of culture.

Stability of tolerance, in the absence of selection pressure, for 1, 2 or 3 passages was examined. Growth after absence of NaCl was compared with growth in the continuous presence of NaCl. Two of the selected lines, R-10 (12 kR) and R-14 (16 kR), proved to be completely stable in tolerance, as measured by growth, when recultured on NaCl (5 or 10 g/l) following culture in the absence of NaCl (Table II). Other lines showed a tendency to reduced growth when grown in the presence of NaCl following absence from it for three subculture periods. This could indicate a possible loss of stability with time but, on the other hand, could well be within the limits of insignificant fluctuations. This later assumption is substantiated by good growth of all lines in the presence of NaCl for several passages after this stability test was carried out.

Tests are underway to determine whether tolerance of lines is ion specific, or based on general tolerance to increased osmoticum. Preliminary results indicate tolerance specifically to Na^+ and Cl^-, a great sensitivity to K^+, irrespective of the anion, and independence of osmoticum in the range of salt concentrations tested (100–200 mH).

Callus lines tolerant to 2, 4-D

Three lines were selected from non-irradiated controls. The growth of these lines in the presence of rising 2, 4-D levels was examined. Two lines showed good

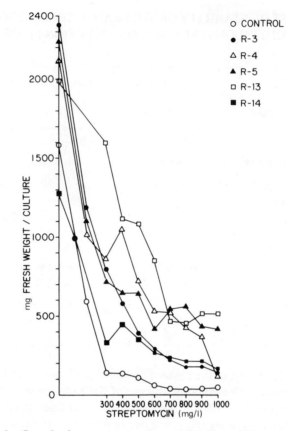

FIG.1.   Growth of streptomycin-tolerant callus lines of Shamouti orange.

tolerance over the whole 2, 4-D concentration range, growing better even on
1×10^{-4} M 2, 4-D than unselected calli on 1×10^{-5} M. Other lines were tolerant
to varying degrees (Table III).

Stability of tolerance, in the absence of selection pressure for 1—3 passages,
was also examined. Only in rare cases was growth reduced after reculture on
2, 4-D following its omission.

Representative results from two lines are presented in Table IV.

The effect of gamma irradiation

Selections for NaCl and 2, 4-D tolerance in Shamouti callus were attempted
with non-irradiated calli and calli irradiated at various doses. Although natural
variation in tissue culture is often considerable [15], irradiation is expected to

TABLE V.   STABILITY TEST FOR STREPTOMYCIN TOLERANCE

| Subculture | Growth on streptomycin sulphate | | | |
| | 175 mg/l | | 400 mg/l | |
	Continuous	Discontinuous	Continuous	Discontinuous
	R-5			
1st	545 ± 75[a]	660 ± 43	405 ± 28	345 ± 49
2nd	800 ± 27	720 ± 43	470 ± 35	435 ± 37
3rd	640 ± 43	680 ± 31	340 ± 28	365 ± 22
	R-14			
1st	615 ± 23	450 ± 33	450 ± 33	300 ± 38
2nd	540 ± 33	665 ± 60	270 ± 27	320 ± 19
3rd	580 ± 50	490 ± 25	360 ± 21	410 ± 23

[a]   mg fresh weight/culture.

increase frequency of mutations and affect the mutation spectrum [16]. As growth under selection pressure was the main parameter in our studies we are not yet able to evaluate fully the effect of irradiation. Studies are underway using suspension cultures to assess the role of irradiation in induction of mutations. On the basis of growth a tendency was evident for a more rapid growth increase during selection passages when selections were started with irradiated callus [8]. The comparison of finally selected lines, however, did not reveal significant differences between lines originating from irradiated and non-irradiated calli. It is yet possible that such differences will be found in plants established from calli.

Callus lines tolerant to streptomycin

Five tolerant lines were selected after recurrent screening of explants cultured in the presence of 175 mg/l streptomycin sulphate. This streptomycin level was found to enable a clear differentiation to be made between susceptible and tolerant lines over prolonged culture periods while inhibition of growth caused by lower streptomycin levels was greatly removed after 10 weeks of culture.

   Growth of selected lines was studied over a broad (175–1000 mg/l) concentration range. Results are presented in Fig. 1, and clearly demonstrate the significantly better growth of all lines on all concentrations tested. Stability of streptomycin tolerance, after absence of selection pressure, was also clearly demonstrated and representative results for two selected lines are given in Table V.

It is postulated that, as streptomycin affects chloroplasts [17], selection for streptomycin tolerance should be carried out with green calli. *Citrus* nucellar callus is white, but the ability to form chloroplasts is evident in green embryos differentiating from the callus. If our selections represent streptomycin-tolerant variants we would expect them to be able to form green embryos in the presence of streptomycin in the medium, as a result of treatments stimulating embryogenesis. Experiments are underway to determine whether streptomycin-tolerant lines are indeed able to form green embryos, or embryos at all.

Resuppression of embryogenesis in selected lines

The final and decisive test is the transfer of the selected trait from the callus line to the plant. *Citrus* nucellar callus is potentially embryogenic and plants can be developed from these embryos. Frequent subculture of callus leads to a reduction in embryogenesis [9]. Whereas NaCl did not affect the embryogenic potential of the callus cultured on saline media, the presence of 2, 4-D in the medium very potently reduced embryogenesis in all but one selected line. A similar effect of reduction in embryogenesis has been observed with other auxin-like growth substances [9, 12].

We established earlier that galactose and other galactose-yielding sugars induce abundant embryogenesis [11]. Callus of selected lines was therefore cultured on media containing 10 g/l galactose or 20 g/l lactose as the only carbon source. Some lines produced embryos at the first subculture passage, others had to be recultured for a second or third passage before embryos appeared.

Treatments which significantly stimulate embryogenesis cause a very high proportion of the embryos (80–100%) to remain small (0.1–0.2 mm) and of globular shape. It is important to further the development of these embryos in order to obtain larger test populations from a selected cell line, instead of testing only the small proportion of embryos which develop spontaneously into more advanced stages. A series of experiments was therefore carried out in order to establish the culture conditions, medium addenda and sequences favourable to the further development of embryos into plants. Based on the results of these experiments the following procedures can now be adopted in order to obtain rooted plantlets in culture:

(a)   Embryos are differentiated from callus on galactose- or lactose-containing media. A solidified medium is best used for this phase. A 10-week culture period yields better results than a shorter culture period.

TABLE VI. COMPARISON OF INHIBITION OF DEVELOPMENT OF
EMBRYOS ORIGINATING FROM NaCl AND 2, 4-D TOLERANT LINES AND
A NON-SELECTED SUSCEPTIBLE LINE[a]

2,4-D (molar)	Relative weight of embryos		NaCl (g/l)	Relative weight of embryos	
	R-1 line	S-line		R-13 line	S-line
0	100[b]	100[b]	0	100[c]	100[c]
1×10^{-7}	120	18	5.0	51	40
1×10^{-6}	81	13	7.5	39	20
1×10^{-5}	116	18	10.0	20	10
5×10^{-5}	91	15	12.5	16	8
1×10^{-4}	53	9	15.0	11	6

[a] Cultured on Petri dishes weighed after 6 weeks.
[b] Average weight 120 mg.
[c] Average weight 312 mg.

(b) These embryos, usually small and globular, are transferred into a suspension culture medium consisting of MS + 20 g/l sucrose, 10 mg/l GA_3 and 1 g/l malt extract for 2–3 weeks, under dim light.

(c) Well-developed cotyledonary embryos are separated from cultures; the remaining small globular embryos can be subcultured again as before. The developed embryos are subsequently transferred to a solid medium, containing MS + 50 mg/l sucrose, 1 mg/l GA_3 and 30 mg/l adenine sulphate for rooting. Dishes should be kept preferably in dim light.

(d) When roots have been initiated, plantlets are transferred to soft agar (0.7%) or alternatively kept on liquid medium with a proper bridge to secure better root development. Sucrose is reduced to 2%.

Level of tolerance of embryos from selected NaCl and 2, 4-D tolerant lines

Globular embryos obtained from NaCl tolerant line R-13 and 2, 4-D tolerant line R-1 were plated on NaCl, and 2, 4-D containing media. Globular embryos from non-selected callus served for comparison. The weight of the embryos is presented in Table VI.

All embryos from NaCl R-13 developed into large cotyledonary embryos and remained green on media containing up to 7.5 g/l NaCl. Embryos from non-selected callus developed into smaller cotyledonary embryos (and were lower in

TABLE VII. ADVENTIVE SHOOT FORMATION ON EMBRYO SHOOT
SEGMENTS IN THE PRESENCE OF SEVERAL BAP LEVELS

BAP level (mg/l)	Shoot forming explants (%)	Shoots per explant (No.)	Frequency (%) of explants forming 1−8 shoots							
			1	2	3	4	5	6	7	8
0	25	1.4	55	45	−	−	−	−	−	−
0.25	52	2.3	32	32	23	3	5	5	−	−
0.50	38	3.2	31	19	12	−	25	7	−	6
1.0	29	2.4	50	14	14	7	7	−	8	−
5.0	30	1.9	44	31	18	7	−	−	−	−
10.0	15	1.0	100	−	−	−	−	−	−	−

weight) but turned brown. On 10 g/l NaCl, 40% of the NaCl R-13 embryos were
green and developed into cotyledonary, somewhat smaller embryos, whereas
control embryos showed very little growth before all turned brown. At
12.5−15.0 g/l NaCl none of the embryos showed viability.

Embryos from 2, 4-D R-1 grew significantly better than control embryos from
non-selected callus, and their weight was considerably larger. All 2, 4-D R-1
embryos remained green up to 5×10^{-5} M 2, 4-D whereas all control embryos
bleached on media containing more than 1×10^{-7} M 2, 4-D. Susceptible embryos
callused. Embryos from 2, 4-D R-1 mostly bleached on 1×10^{-4} M 2, 4-D, but
still 10% of them remained green even on this high level.

An improved tolerance of embryos from selected lines to 2, 4-D or NaCl
respectively was thus demonstrated. This encouraging result will have to be
substantiated by further observations including rooted plants.

Embryo and cell cloning

Selections have so far been made using callus explants. The advantages of
this method are its simplicity and the relative insensitivity of calli to possible side
effects of the selective agent such as increased osmotic pressure in a salt-containing
medium. On the other hand, one must consider that explants from which
selections are started contain about 1×10^5 cells. Not all of them are in direct
contact with the selective medium, and gradients of the selective agent are
expected to exist throughout the cell mass. Cross-feeding could also enable
susceptible cells to maintain divisions. Therefore a series of selection passages
was devised which would result in a substantial increase of the resistant cell population in selected lines.

Although the stability tests, on the basis of growth, were satisfactory, one cannot assume that all cells represent the resistant or highly tolerant type. Embryos formed will reflect at best the proportions of the various cell types present. A possibility exists for preferential embryo formation from one or other cell type, and shifts in the proportion of cell types might occur when selection pressure is removed during resuppression of embryogenesis.

The major problem in assessing plants from selected cultures is their possible genetic diversity. The size of the tested plantlet may also have an effect on the tolerance level exhibited. It is therefore useful to regard each embryo as an independent unit unless uniformity of the selected line is proven. Even replicated testing involving different embryos (or plantlets) from the same selected line does not in itself guarantee adequate uniformity and repeatability.

To overcome this problem studies were initiated in two directions:

(a)   to obtain clonal multiplication from individual embryos, which would enable the replicated testing of each embryo

(b)   to develop a selection system based on suspension cultures combined with plating of single cells or small cell groups which can be considered as originating from a single cell.

Clonal multiplication via adventive shoot formation

A series of experiments was performed to obtain adventive shoot formation from embryonal shoot segments. Segments were either very short fasciated shoot tips, often observed when plants are developed from embryos, or segments from normally elongated shoots from plantlets originating from embryos.

The experimental approach was based on a report by Chaturvedi and Mitra [18] who obtained adventive shoots from seedling stem segments on a medium containing malt extract and different combinations of NAA and BAP.

Segments, whether derived from short fasciated tips or from elongated normal shoots, formed shoots in all combinations tested (Table VII). The greatest number of shoots/explant was obtained combining BAP 0.5 mg/l + NAA 0.1 mg/l, but somewhat lower or higher BAP levels (0.25 mg/l or 0.1 mg/l) gave similar results.

Explants on which shoots were formed were recultured for a second time after removal of shoots. Shoot formation increased to an average of 13.4 shoots per explant. The origin of shoots (whether from single cells or multicellular buds on the segments) is not yet known. Although of basic interest, this has no practical significance under the scope of these studies.

Experiments are underway to obtain rooting of the detached shoots. Preliminary results indicate that certain NAA levels promote root formation. Rooted shoots, once obtained in sufficient numbers from embryos of tolerant lines, will then enable replicated testing to be made for the expression of tolerance in plants.

Suspension culture and plating of cell groups

Suspension cultures were initiated from nucellar callus of all *Citrus* cultivars at present available. After initial difficulties caused by the tendency of the tissue to form clumps, we have obtained so far finely dispersed suspensions from Shamouti orange and Villafranca lemon. Studies were initiated to learn the kinetics of *Citrus* suspension cultures and to establish minimal plating densities necessary for colony formation. Since we started this work only recently we can comment only on preliminary observations. Growth in suspension culture is more rapid than on solid media. Maximal growth was obtained after 12—16 days, within which cells underwent 5—6 cell cycles. The exponential phase of growth is between the 3rd and 12th day of culture. In successful plating experiments we obtained 35—40% of plating efficiency with an inoculum density of 360—460 cell groups/m. These groups were smaller than 280 μm and had 16—32 cells.

Our work is far from being completed. Great emphasis will be given to testing plants regenerated from selected tolerant calli in order to find out if NaCl, 2, 4-D and streptomycin tolerance, observed in the cell culture phase, is transferable to plants. Additional selections will be carried out based on embryogenic calli from other *Citrus* species. Suspension culture methods will be employed to an increasing extent. This should enable us to evaluate the role of ionizing radiation in the induction of mutations and recovery of mutants.

ACKNOWLEDGEMENTS

The valuable assistance of S. Saad and H. Neuman is gratefully acknowledged. We wish to thank J. Button and A. Vardi for nucellar cultures of several varieties.

REFERENCES

[1] DONNINI, B., "The use of radiations to induce useful imitations in fruit trees", Research Co-ordination Meeting Wageningen, 1976, Tech. Doc. IAEA-194, 55 (unpublished).
[2] LAPINS, K.O., "Induced mutations in fruit trees", Induced Mutations in Vegetatively Propagated Plants, (Proc. Panel Vienna, 1972), IAEA, Vienna (1973) 1.
[3] BOTTINO, P.J., Radiat. Bot. **15** (1975) 1.
[4] MALIGA, P., in Cell Genetics in Higher Plants (DUDITS, D., FARKAS, G.L., MALIGA, P., Eds), Hung. Acad. Sci. (1976) 59.
[5] MALIGA, P., in Frontiers of Plant Tissue Culture (Proc. 4th Int. Congress Plants Tissue and Cell Culture, Calgary 1978) (THORPE, T.A., Ed.).
[6] MELCHERS, G., Z. Pflanzenzuecht. **67** (1972) 19.
[7] THOMAS, E., KING, P.J., POTRYKUS, I., Z. Pflanzenzuecht. **82** (1979) 1.

[8] KOCHBA, J., SPIEGEL-ROY, P., SAAD, S., paper presented at IAEA Research Co-ordination Meeting Skiernewitze, 1978 (unpublished).

[9] KOCHBA, J., SPIEGEL-ROY, P., Z. Pflanzenphysiol. **81** (1977) 283.

[10] KOCHBA, J., SPIEGEL-ROY, P., Environ. Exp. Bot. **17** (1977) 151.

[11] KOCHBA, J., SPIEGEL-ROY, P., SAAD, S., NEUMANN, H., Naturwissenschaften **65** (1978) 261.

[12] KOCHBA, J., SPIEGEL-ROY, P., NEUMANN, H., SAAD, S., Z. Pflanzenphysiol. **89** (1978) 427.

[13] KOCHBA, J., SPIEGEL-ROY, P., SAAD, S., in Plant Cell Cultures, Results and Perspectives, Proc. Int. Workshop Pavia, Elsevier, Amsterdam (1980).

[14] SPIEGEL-ROY, P., KOCHBA, J., in Advances in Biochemical Engineering **16**, Springer Verlag, Berlin (1980) 27.

[15] SKIRVIN, R.M., Euphytica **27** (1975) 241.

[16] HOWLAND, G.P., HART, R.W., in Plant Cell, Tissue and Organ Culture, (REINERT, J., BAJAJ, Y.P.S., Eds), Springer Verlag, Berlin (1977) 731.

[17] UMIEL, N., GOLDNER, R., Protoplasma **89** (1976) 83.

[18] CHATURVEDI, H.C., MITRA, G.C., Hortscience **9** (1974) 118.

INDUCTION OF MUTATIONS IN CITRUS
FOR THE DEVELOPMENT OF RESISTANCE TO
Xanthomonas citri (Hasse) Dowson*

H.M. ZUBRZYCKI, A. DIAMANTE DE ZUBRZYCKI
Estación Experimental Agropecuaria INTA,
Corrientes,
Argentina

Abstract

INDUCTION OF MUTATIONS IN CITRUS FOR THE DEVELOPMENT OF RESISTANCE
TO *Xanthomonas citri* (Hasse) Dowson.

With the aim of obtaining resistance to *Xanthomonas citri* in citrus, mutagenic treatment
was carried out with X-rays and gamma rays on buds and seedlings. As a prerequisite, attempts
were made to determine some physiological and structural differences in leaves in order to
evaluate and analyse the reaction to the bacterium in qualitative and quantitative terms. Citrus
in Bella Vista, Corrientes, Argentina, were found to present 3—4 annual sproutings, each having
a different intensity. The growth of autumn sprouts is more marked in a cultivar susceptible
to *Xanthomonas citri* (grapefruit) than in the more resistant Valencia Wood (orange). Grapefruit
showed a greater initial increase and final length than orange, but lemon growth was greater
than all the others. Valencia is the first to stop growing. According to their growth, leaves
clustered in three groups. Those in the middle part of the sprout showed the greatest daily
increase, followed in decreasing order by those of the base and those of the apex. In orange
cultivars with different degrees of reaction to *Xanthomonas citri* a direct association between
daily growth of leaves and susceptibility was found. In fully developed leaves the number of
stomata per surface unit presented an inverse ratio with susceptibility. It was observed that
in a given genotype the stomata frequency per surface unit was higher in young leaves than in
developed ones. Also a direct association between the leaf area and the number of stomata
per leaf was found. Natural field infections were measured in leaves of developed plants.
Infection in each of the 3—4 annual sproutings depends on environmental factors during the
growth period and the predisposition period of the leaves. The available variation for reaction
to *Xanthomonas citri* was higher in oranges than in grapefruit. The abscission of each leaf
depends on its degree of infection, but the cultivars classed as more resistant are those
requiring a smaller affected area to provoke the leaf abscission. However, abscission of a leaf
is independent of the fate of the neighbouring ones. The existence of a continuous grading of
infection damage within the orange cultivars suggests a genetic component for resistance to
Xanthomonas citri.

INTRODUCTION

Citrus canker, a disease caused by *Xanthomonas citri,* has been affecting
citrus plantations in South America for a number of years. This bacterium affects
the most important commercial species, and can cause dramatic and fundamental

---

* Research supported by the IAEA under Research Contract No. 1731.

changes in the American citrus economy because of its symptomatology, aggressiveness and virulence.

Early in the 1970s Argentine citrus plantations were markedly expanding, a trend that seemed irreversible. Half-way through the decade, however, cancrosis was found in some minor citrus areas in the Eastern provinces and then spread to affect several Argentine provinces. This situation has caused a recess in citrus expansion and plantations have been abandoned or reduced in size.

Cancrosis has been reported to affect lemon trees in Argentina since as early as 1928 but it was only in 1973–75 that it reached an epidemic level in the east, affecting other species such as oranges, grapefruit, tangerines, lime, etc.

Some reports contend that this bacterium, which attacks stems, leaves and young fruit in active growth, gains entrance through stomata and lenticels [1]. Other works report on the humidity and temperature required for infection and disease development [2].

McLean [3] stated that resistance to *Xanthomonas citri* is determined by the structure of stomata.

Stall and co-workers [4] observed a new type of resistance in the citrus leaves determined by the mesophyll cells. This type of resistance was thought not to apply to *Xanthomonas citri* because tissue injury nullifies the resistance.

Many workers have reported that resistance to *Xanthomonas citri* varies with species, and variety variability in resistance to *Xanthomonas citri* has been observed; no observations have been published, however, on the predisposition of different species to the pathogen.

Leaves go through different stages of susceptibility to *Xanthomonas citri* during their growth. The speed of this maturity process could depend on the genotype or on environmental factors.

Understanding of these aspects would allow the determination of different mutants and the selection of the most suitable ones for resistance.

The artificial induction of mutations in vegetatively propagated plants can be brought about by breeding as has been demonstrated by Nybom [5] and Nybom and Koch [6], and by International Atomic Energy Agency in its specific programme [7, 8].

The work reported here involved the use of X-rays and gamma rays on citrus seedlings and seeds to obtain resistant mutations to *Xanthomonas citri*. As a prerequisite, efforts were made to determine some physiological and structural differences in leaves in order to evaluate and analyse the reaction to the bacterium in qualitative and quantitative terms.

MATERIALS AND METHODS

Grapefruit plants treated with various doses of X-rays and gamma rays were used in this work. The grapefruit varieties were: Thomson, Ruby, Foster, Marsh

Seedless, Excelsior, Meteor, Florida, Pernambuco, California, McCarthy and Duncan.

Nucellar seedlings were exposed to gamma rays from a ^{60}Co source for periods of one and two years. After this treatment, buds V_1 were grafted to Cleopatra *(Citrus reshni* Hort) and Rangpur *(Citrus limonia,* Osb.) rootstocks.

Seeds were X-ray treated under 230 kV/15mA with 0.10 kr and 30 kr doses. When they had grown, the M_1 seedlings were transplanted to the field.

Between 1966 and 1967 all treated plants were transplanted to the field in a 7m X 7m design. Each treatment was carried out on five to nine plants.

All the material was subjected to natural infection since it was planted near other heavily infected plantations.

Natural field infections on leaves of developed trees were evaluated on treated plants as well as in 39 orange and five lemon cultivars during four consecutive periods. Infection was estimated in samples of 100 leaves of the same growth period taken at random from all corners of the tree. The average leaf area was determined for each sample. The samples were grouped into five categories depending on the number of lesions per leaf, as follows:

0 = no lesion per leaf
2 = 1–4 lesions per leaf
3 = 5–9 lesions per leaf
4 = 10–21 lesions per leaf
5 = over 21 lesions per leaf.

Based on these categories, damage intensity (DI) was evaluated according to methods described in Trans. Br. Mycol. Soc. 31 (1948) 343.

The number of lesions per cm^2 of leaf surface was determined by taking into account the number of lesions and the average leaf area.

In order to establish the relation between affected leaf area and leaf abscissions, fallen leaves were collected, and total leaf area, number of lesions and area with yellow halo (or affected area) were determined. The percentage of affected area was then estimated in relation to the total leaf area.

Furthermore, artificial infection of the mesophyll cells was used to evaluate the non-treated material and performed by the injection method as presented by Stall and co-workers [4]. Groups of three leaves taken from three different shoots of the same age per cultivar were inoculated with a concentration of 5 X 10^3 cells/ml of *Xanthomonas citri.* After 20 days the infection was evaluated by counting the number of lesions per cm^2 of leaf surface. The reaction of each cultivar was determined by the average infection values.

In the region where this evaluation was carried out, *Xanthomonas citri* (Hasse) Dowson is evenly distributed at an epidemic level.

The irradiated material was not sprayed with chemicals. On the contrary, orange and lemon cultivars were periodically sprayed with copper.

The frequency and intensity of sprouting were measured on the above species and cultivars for eight years. Using a five-step arbitrary scale, sprouting intensity was measured by the following equation

$$\text{Sprouting intensity} = \frac{(1 \times N_1) + \ldots\ldots (5 \times N_5)}{5\,N} \times 100$$

where $1 = 20\%$; $2 = 40\%$; $3 = 60\%$; $4 = 80\%$ and $5 = 100\%$ of the tree sprouted surface

$N_1 \ldots\ldots\ldots N_5$ = number of individuals per step
N = total number of individuals.

Shoots were classified as follows: (1) small, (2) medium and (3) large.

The growth of shoots and leaves was measured in two trees per cultivar and in 16 shoots per tree. Weekly measurements were carried out throughout the growth period.

Leaf area was determined by measuring the maximum width and the length of each leaf of the shoot.

Stomata density was established in five trees per cultivar and in 16 leaves per tree. Leaves in different growth stages from very young to fully developed were sampled from the middle of the shoots, on the four opposite sides of the tree.

The lower surface of the leaves was covered with a vinylic adhesive. When dry, the adhesive was removed from the leaf surface and the number of stomata per mm^2 of leaf surface was then determined by observing the dry adhesive under the microscope with a 40/0.65 D and micrometric ocular.

Nujol oil was used to determine stomata opening and closure times since only open stomata absorb this oil which forms oily stains on the leaf surface.

All the above materials are maintained at INTA Bella Vista, Corrientes, Argentina, a station located 70 m above sea level, latitude 28°26' S, longitude 58°55' W, on sandy soil. Annual rainfall is 1.200 mm; average relative humidity, 72%; maximum temperature, 33°C; minimum temperature, 8.5°C; and average monthly temperature, 20.5°C (data supplied by the Agricultural Meteorology Service, INTA, Bella Vista, Corrientes).

Data were analysed and evaluated by linear regression, Wilcoxon's test of rank addition and Hotelling and Pabst's test of rank correlation.

RESULTS AND DISCUSSION

In Bella Vista, Corrientes, citrus trees present 3–4 annual sprouts in spring, summer and autumn respectively. Each sprouting has a different intensity, the

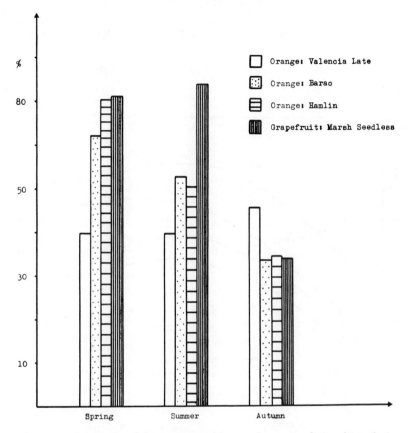

FIG.1. *Intensity of sprouting of three orange cultivars and one grapefruit cultivar during different seasons; eight years average value.*

spring one being the most intense of all. It was observed that sprouting intensity varies not only with the time of the year but also with cultivars (Fig.1).

Figure 2 shows sprouting growth estimated according to a quadratic function. Marsh seedless grapefruit and Petrópolis orange grow larger and for a longer period than the other varieties. Valencia Wood orange presents the poorest and shortest growth.

There is a direct relation between growth and reaction of susceptibility to *Xanthomonoas citri*. Valencia Wood is the most resistant variety and Marsh Seedless and Petrópolis are the most susceptible ones.

Leaves were clustered in three groups depending on their daily growth in autumn: the leaves in the middle of the shoot (Group II) showed the greatest daily growth, followed in a decreasing order by those of the base (Group I) and then by those of the apex (Group III) (Fig.3).

(text cont. on p.100)

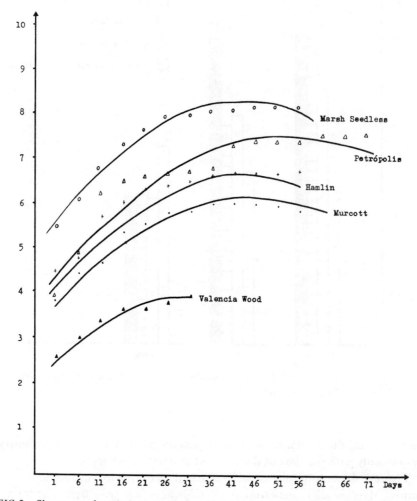

FIG.2. Shoot growth variation against time (days) in five citrus cultivars of different species.

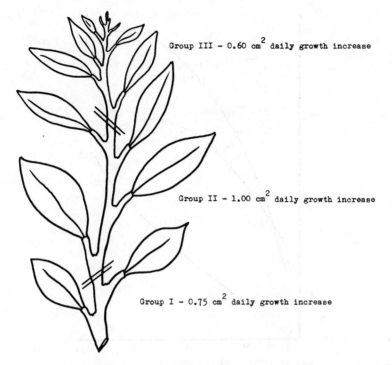

Group III – 0.60 cm^2 daily growth increase

Group II – 1.00 cm^2 daily growth increase

Group I – 0.75 cm^2 daily growth increase

FIG.3. Banking of leaves in a shoot according to their daily growth increase. Autumn flush of growth was evaluated. The average of growth increase values was found in four orange cultivars of different maturity times.

TABLE I. DAILY GROWTH INCREASE IN THREE TYPES OF LEAVES IN AUTUMN SPROUTS OF FOUR ORANGE CULTIVARS WITH DIFFERENT REACTIONS OF SUSCEPTIBILITY TO *Xanthomonas citri*

Cultivars	Leaves Group I (cm^2/day)	Leaves Group II (cm^2/day)	Leaves Group III (cm^2/day)	Susceptibility to *Xanthomonas citri*
Petrópolis	0.96	1.38	1.05	Highly susceptible
Hamlin	0.83	1.09	0.58	Slightly susceptible
Criolla	0.64	0.90	0.52	Moderately susceptible
Valencia Wood	0.60	0.64	0.28	Moderately resistant

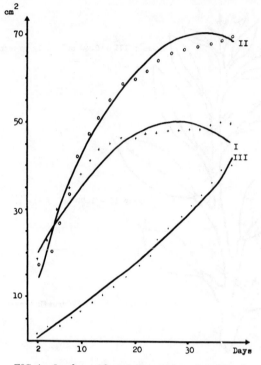

FIG.4.   Leaf growth variation against time (days)
in the three groups of leaves of Petrópolis
orange shoots.

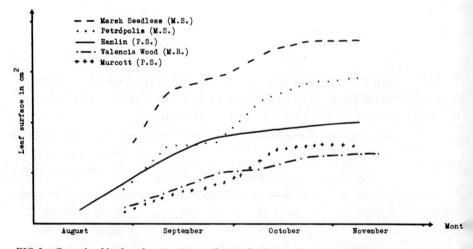

FIG.5.   Growth of leaf surface in winter. Spring flushes in five citrus cultivars with different
reactions of susceptibility to Xanthomonas citri. M.S. = very susceptible; P.S. = slightly
susceptible; M.R. = moderately resistant.

TABLE II. STOMATIC DENSITY PER SURFACE UNIT AND REACTION OF
SUSCEPTIBILITY TO *Xanthomonas citri* IN DIFFERENT CITRUS CULTIVARS

Citrus cultivars	Stomata/mm^2 $\bar{x}$	Susceptibility to *Xanthomonas citri*
Hamlin	513 a*	Slightly susceptible
Sucral Vive	508 b	Moderately resistant
Enterprise	457 c	Moderately susceptible
Marrs Early	377 d	Highly susceptible
Sanguinelli Vil	473 a	Moderately resistant
Petrópolis	429 b	Highly susceptible
Valencia Campbell	434 a	Highly resistant
Valencia Wood	410 b	Moderately resistant
Natal San Pablo	401 c	Moderately susceptible

* Values followed by different letters differ significantly between each other (Wilcoxon's test
of rank addition).

TABLE III. AVERAGE VALUE OF NUMBER OF STOMATA PER SURFACE
UNIT IN CITRUS LEAVES AT DIFFERENT GROWTH STAGES AND
BELONGING TO YOUNG AND ADULT PLANTS

	Young plants		Adult plants	
	Leaf surface (cm^2)	Number of stomata (mm^2)	Leaf surface (cm^2)	Number of stomata (mm^2)
Orange Valencia Late	6.2	1105	5.9	515
	19.3	694	39.3	628
	43.2	529	50.1	470
Lemon Villafranca	7.6	642	6.5	180
	44.6	644	41.1	316
	69.6	667	51.3	308
Grapefruit Marsh Seedless	12.9	885	8.1	497
	47.0	544	36.1	702
	49.3	507	55.8	600

TABLE IV. LEAF SURFACE, NUMBER OF STOMATA $\times$ 10^{-3}, REGRESSION
BETWEEN LEAF SURFACE AND NUMBER OF STOMATA AND ITS
DETERMINANT COEFFICIENT (r^2) OF ORANGE AND GRAPEFRUIT SPECIES

Citrus species	Leaf surface $\bar{x}$ (cm^2)	Number of stomata/leaf $\times$ 10^{-3} $\bar{x}$	Regression (leaf surface/ number of stomata)	r^2 (%)
Orange (Valencia Wood)	5.5	335	0.97	86
	39.3	2479	0.47	46
	50.6	2364	0.49	61
Grapefruit (Marsh Seedless)	8.1	453	0.76	46
	36.0	2538	0.69	74
	55.8	3389	0.78	88

Table I shows the daily growth of these three groups of leaves in the same
shoot in four orange cultivars. A direct relation between daily growth of leaves
and reaction of susceptibility to *Xanthomonas citri* was determined by considering these values of daily growth and the different degrees of susceptibility
per cultivar.

Leaf growth was estimated as a function of time in the three groups of
leaves in shoots of all the cultivars studied. Taking Petrópolis orange as an example,
Fig.4 shows the difference in growth among each group of leaves. Their growth
was estimated through a quadratic equation.

Leaf growth was determined in spring-winter sprouts and it was observed
that it varied with species and cultivars. A direct relation between greater leaf
growth and the reaction of susceptibility of the cultivar was determined (Fig.5).

Since stomata are natural entrance sites for *Xanthomonas citri,* it was
important to determine if the stomata density per surface unit had any relation
with cultivars having different degrees of resistance or susceptibility to this
bacterium. In fully developed leaves of different orange varieties, the number
of stomata per mm^2 of leaf surface presented an inverse ratio to the reaction of
susceptibility to *Xanthomonas citri* (Table II). This inverse reaction shows that
stomata density might not determine the difference in resistance to natural
infection, but as stated by McLean [3], the converse would be true for other
factors such as number or shape of stomata.

The number of stomata per unit of leaf surface was analysed in leaves in
different growth stages which had been collected from trees of different ages.
Higher stomata density was observed in younger plants (Table III).

A statistically significant direct relation between the number of stomata per leaf and leaf area was found. The determinant coefficient r^2 was also significant (Table IV).

In relation to the activity of the stomata in orange and grapefruit, it was observed that in the autumn opening takes place at about 07:30, and closing at about 18:00 (sunrise and sunset respectively).

It was observed that infection in each of the 3–4 annual sproutings depends on environmental factors during the growth period and the predisposition period of the leaves. According to the DI values shown in Table V, it was determined that higher temperature corresponded to a higher degree of infection. This relation was not observed for humidity, rain and dew but, since their values are high, it is thought that they might favour infection. Because of these factors, infection evaluation must be carried out independently in each sprout, but the best quantification occurs in spring and summer.

On the basis of the number of lesions per square centimetre of leaf surface, it was established that the early, intermediate and late orange cultivars belong to the same population in terms of their host-pathogen reaction (Fig.6). Differences in reaction, however, exist among the three groups, and in Valencia Late there are even remarkable differences between clones (Table VI).

When the infection values in orange and grapefruit varieties naturally infected in the field were examined, it was observed that the available variation of reaction to *Xanthomonas citri* was higher in orange than in grapefruit (Tables VII and VIII respectively).

The continuous grading of infection present in orange cultivars suggests the presence of genetic components for reaction, resistance or susceptibility to *Xanthomonas citri* (Table VII).

It was concluded that early abscission of infected leaves depends on the degree of infection and that the most resistant varieties (smaller number of lesions per cm^2 of leaf surface) require a smaller affected area to cause leaf abscission (Fig.7 and Table IX). On the other hand, the abscission of a leaf in a shoot is independent of the fate of the neighbouring ones.

It was considered that an artificial infection might speed up the study of resistance or susceptibility to *Xanthomonas citri,* and the number of lesions caused by natural infection per cm^2 of leaf surface was therefore compared with those caused by inoculum injection in the mesophyll cells. Table X shows that there is no correspondence between the values obtained by both methods, except for Lima Key which presents the highest values in both cases.

This fact suggests that there could be two independent types of reaction of resistance to *Xanthomonas citri*, one resistant to infection at an external level and the other resistant at an internal level or in the mesophyll cells. Should this be the case, it would be very important to combine the two types of reaction in the same genotype but this possibility has not yet been achieved. As it is, the reaction to natural infection would seem more useful.

TABLE V. AVERAGE VALUES OF DAMAGE INTENSITY (DI) CAUSED BY *Xanthomonas citri* IN LEAVES OF 55 CULTIVARS OF DIFFERENT CITRUS SPECIES: THEIR POSSIBLE RELATION TO TEMPERATURE HUMIDITY, RAIN AND DEW

Data on climate are average values of the month except those on rain and dew which are totals

Evaluated period of sprouting and infection	DI Average species	°C $\bar{x}$ monthly	°C $\bar{x}$ minimum	°C $\bar{x}$ maximum	Relative humidity (%)	Rain (mm)	Total dew hours
October–November 1978	5.1	21.9	17.7	27.1	76	425	359
February–March 1979	11.7	23.7	19.2	29.6	76	251	279
December–January 1979 1980	14.3	25.5	20.2	31.1	71	228	646

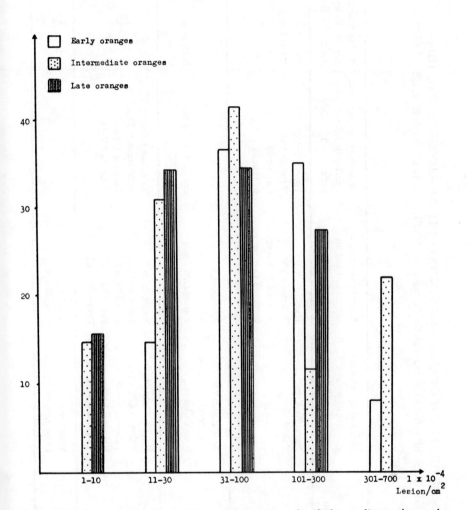

FIG.6.   *Frequency of early, intermediate and late oranges classified according to the number of lesions per cm² of leaf surface.*

TABLE VI. CLASSIFICATION OF DIFFERENT ORANGE AND LEMON CULTIVARS ACCORDING TO THEIR DEGREES OF RESISTANCE OR SUSCEPTIBILITY TO *Xanthomonas citri*, EXPRESSED BY NUMBER OF LESIONS PER cm² OF LEAF SURFACE ($\times 10^{-4}$)

Average values 1977–1979

Cultivars	Highly resistant: number of lesions per cm² = 1–10	Moderately resistant: number of lesions per cm² = 11–30	Slightly susceptible: number of lesions per cm² = 31–100	Moderately susceptible: number of lesions per cm² = 101–300	Highly susceptible: number of lesions per cm² = 301–700
Early Orange	Cocco Salustiana	Navelina Vive Sucral Vive	Hamlin Thomson 390 Atwood E. Navel Ombligo Lentijo Westin	Trovita USA Navelate Sar Enterprise Bahianina	Marrs Early
Intermediate Orange		Sanguinelli Vil Cadenera Naranja Lima	Pineapple Barao Naranja del Cielo Criolla Mejorada	Criolla Temprana	Petrópolis China
Late	Valencia Campbell	Hughes Valencia Valencia LPC 466 Valencia Wood Valencia Late Valencia CN 35	Valencia sp. Lue Gim Gong Pera 1743–82 Pera 1743–27 Pera 1743–52	Natal San Pablo Natal Nucelar Brasil Valencia Frost Valencia CN 44	
Lemon	Seedless	Villafranca Frost Lisbon	Eureka Frost Eureka		

TABLE VII. CLASSIFICATION OF ORANGE CULTIVARS ACCORDING
TO THE NUMBER OF LESIONS CAUSED BY *Xanthomonas citri* PER cm^2
OF LEAF SURFACE
Average value 1977–79

Orange cultivars	1×10^{-4} lesions per cm^2	Orange cultivars	1×10^{-4} lesions per cm^2
Cocco	4	Criolla Mejorada	60
Valencia Campbell	7	Thomson 390	62
Salustiana	8	Lue Gim Gong	65
Sanguinelli Vill	11	Westin	65
Navelina Vive	12	Ombligo Lentijo	66
Valencia Late (B.V.)	13	Pera 1743–52	90
Valencia Wood	14	Atwood Early Navel	98
Sucral Vive	14	Natal San Pablo	134
Hughes Valencia	18	Trovita USA	150
Cadenera	20	Enterprise	160
Valencia LPC 466	22	Natal Nucelar Brasil	169
Naranja Lima	23	Navelate Sar 127	171
Valencia CN 35	29	Valencia Frost	177
Barao	45	Valencia CN 44	177
Pera 1743–27	46	Criolla Temprana	181
Hamlin	52	Bahianina	266
Valencia sp.	57	Marrs Early	469
Pera 1743–82	57	China	650
Naranja del Cielo	59	Petrópolis	667
Pineapple	59		

The solution of the problem through the use of mutations seems highly promising. So far four putative mutants resistant to natural infections of *Xanthomonas citri* have been analysed in grapefruit, one of which originated spontaneously and three were induced through X-ray treatment (Table XI). This table shows that the number of lesions per cm^2 of leaf surface is many times lower in mutants than in their controls. Apart from this, these mutants present other differences in plant and leaf size although in other characteristics they resemble their original clones.

TABLE VIII. CLASSIFICATION OF GRAPEFRUIT
CULTIVARS ACCORDING TO THE NUMBER OF
LESIONS CAUSED BY *Xanthomonas citri* PER cm^2
OF LEAF SURFACE
Average value 1978—79

Grapefruit cultivars	Number of lesions per cm^2 of leaf surface
Florida	848
Meteor	1052
Marsh Seedless	1158
California	1201
McCarthy	1202
Foster	1246
Pernambuco	1264
Thomson	1296
Duncan	1338
Ruby	1411
Excelsior	1524

This material was observed during four sproutings in two consecutive years, and will be further observed for three additional years before the reaction is finally established.

SUMMARY AND CONCLUSIONS

It was established that citruses in Bella Vista, Corrientes and Argentina, present 3—4 annual sproutings, each having a different intensity. The intensity of each sprouting of the various species and cultivars was studied and their characteristics noted.

It was observed that the growth of autumn sprouts is more marked in a cultivar susceptible to *Xanthomonas citri* (grapefruit) than in the more resistant Valencia Wood (orange). Grapefruit spring sproutings show a greater initial increase and final length than orange, but lemon growth is greater than all the others. Among oranges, Valencia is the first to stop growing.

In terms of growth leaves clustered in three groups. Those in the middle part of the sprout (Group II) showed the greatest daily increase in size followed in

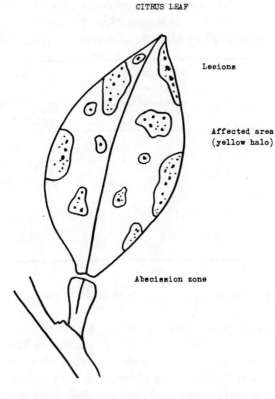

FIG.7. *Scheme of a citrus leaf showing the lesions*
and yellow halo caused by Xanthomonas citri.

TABLE IX. RATIO BETWEEN BLADE SURFACE, NUMBER OF LESIONS
AND AREA AFFECTED BY *Xanthomonas citri* IN LEAVES FALLEN FROM
DIFFERENT CITRUS SPECIES AND CULTIVARS

Cultivars	Lamina area (cm^2)	Number of lesions per leaf	Number of lesions (cm^2)	Affected area of lamina (%)
Petrópolis	54.7	159	2.9	58.5
Lima Key	13.9	36	2.6	44.2
Marsh Seedless	39.2	84	2.1	34.7
Valencia Late	48.4	39	0.8	20.7

TABLE X. RELATION BETWEEN REACTION OF RESISTANCE TO
NATURAL INFECTION IN LEAVES AND RESISTANCE IN MESOPHYLL
CELLS IN LEAVES OF DIFFERENT CITRUS SPECIES AND CULTIVARS

Species and cultivars	Inoculation by injection method: number of lesions/cm^2	Natural infection in field: number of lesions/cm^2 leaf surface
Orange Marrs Early	6.7	0.0469
Orange Petrópolis	10.2	0.0667
Orange Cocco	11.7	0.0004
Orange Valencia Campbell	8.1	0.0007
Lima Key	132.0	0.1630
Lima Tahiti	20.5	0.0003

decreasing order by those of the base and those of the apex (Groups I and III respectively).

A direct association between daily growth of leaves and susceptibility was found in orange cultivars with different degrees of reaction to *Xanthomonas citri.* In fully developed leaves, in turn, the number of stomata per surface unit presents an inverse ratio to susceptibility. It was observed that in a given genotype the stomata frequency per surface unit was higher in young leaves than in developed ones. Furthermore, a direct association between leaf area and the number of stomata per leaf was found.

In relation to the activity of the stomata in orange and grapefruit it was observed that opening takes place at 07:30–08:00 and closing at 17:30–18:00.

Natural field infections were measured in leaves of developed plants of 39 orange, 5 lemon and 11 grapefruit cultivars for three consecutive years. Infection was determined by the number of lesions per cm^2 of leaf surface and damage intensity (DI) was observed on five plants of each cultivar.

It was noted that the infection in each of the 3 or 4 annual sproutings depends on environmental factors during the growth period and the predisposition of the leaves. Because of these facts individual evaluations must be carried out on each sprouting, although the best quantification of infection occurs always in spring and summer.

On the basis of the number of lesions per cm^2 of leaf surface, values of the early, intermediate and late orange cultivars belong to the same population in respect of their reaction. However, differences in reaction exist within the three groups. In the Valencia Late cultivar there are even remarkable differences between clones.

TABLE XI. SOME CHARACTERISTICS OF PUTATIVE MUTANTS OF
GRAPEFRUIT WITH REACTION OF SLIGHTER SUSCEPTIBILITY TO
Xanthomonas citri IN RELATION TO THEIR RESPECTIVE CONTROLS
Average of two years and two observations per year

Cultivar	Treatment	Denomination of putative mutant	1×10^{-4} lesions per cm^2 of leaf surface	Plant area (m^2)	Leaf area (cm^2)
Marsh	Control	—	1.158	10.0	39.9
Seedless	30 kR	No.49	265	10.2	45.1
McCarthy	Control	—	1.202	9.8	51.8
	10 kR	No.138	211	5.6	38.0
Excelsior	Control	—	1.524	13.3	50.5
	10 kR	No.72	415	6.2	49.4
Duncan	Control	—	1.338	12.6	48.9
	0	No.253	491	9.2	48.1

The available variation for reaction to *Xanthomonas citri* was higher in oranges than in grapefruit.

It was concluded that the abscission of each leaf depends on its degree of infection but the cultivars defined as more resistant are those requiring a smaller affected area to cause leaf abscission. On the other hand, abscission of a leaf is independent of the fate of the neighbouring ones.

The existence of a continuous grading of infection damage within orange cultivars suggests a genetic component for resistance to *Xanthomonas citri.*

The data on natural and artificial infection suggest that there might be two independent types of reaction of resistance to *Xanthomonas citri:* one would be resistance to natural infection at an external level, and the other resistance at an internal level or in the mesophyll cells.

So far four putative mutants resistant to *Xanthomonas citri* have been found in grapefruit, one of which was spontaneous and three originating from X-ray treatments. They will be under close observation for a period not shorter than five years.

ACKNOWLEDGEMENTS

The authors are grateful for the financial support received from the IAEA. They also express their thanks to E.A. Favret for reading and discussing the paper, and to B.I. Cantero de Echenique for preparing the inoculum and infecting the mesophyll cells artificially.

REFERENCES

[1] PELTIER, G.L., FREDERICH, W.J., Relative susceptibility to citrus canker of different species and hybrids of the genus citrus, including the wild relatives, J. Agric. Res. **XIX** 8 (1920) 339.

[2] PELTIER, G.L., Influence of temperature and humidity on the growth of *Pseudomonas citri* and its host plants and on infection and development of the disease, J. Agric. Res. **XX** 6 (1920) 447.

[3] McLEAN, F.T., "A study of the structure of the stomata of two species of citrus in relation to citrus canker", Plant Pathology 1 (HORSFALL, J.G., DIMOND, A.E.,Eds), Academic Press, New York and London (1950) 397.

[4] STALL, R.E., CANTEROS DE ECHENIQUE, B.I., MARCO, G.M., ZUBRZYCKI, H.M., ⟪Resistencia a *Xanthomonas citri* (Hasse) Dowson en hojas de citrus de diferentes variedades⟫, Cancrosis de los cítrus, Proyecto Cooperativo INTA-IFAS, Informe técnico N°1, Capítulo III (1979).

[5] NYBOM, N., "The use of induced mutations for the improvement of vegetatively propagated plants", Mutation and Plant Breeding, Rep. NASN-RC 891 (1961) 252.

[6] NYBOM, N., KOCH, A., "Induced mutations and breeding methods in vegetatively propagated plants", The Use of Induced Mutations in Plant Breeding (Rep. FAO/IAEA Tech. Meeting Rome, 1964), Pergamon Press, Oxford (1965) 659.

[7] INTERNATIONAL ATOMIC ENERGY AGENCY, Improvement of Vegetatively Propagated Plants through Induced Mutations (Proc. Research Co-ordination Meeting Tokai, 1974), Tech. Doc. No.173, IAEA, Vienna (1975) (unpublished).

[8] INTERNATIONAL ATOMIC ENERGY AGENCY, Improvement of Vegetatively Propagated Plants and Tree Crops through Induced Mutations (Proc. 2nd Research Co-ordination Meeting Wageningen, 1976), Tech. Doc. No.194, IAEA, Vienna (1976) (unpublished).

INDUCTION OF MUTATIONS IN CITRUS TO PRODUCE RESISTANCE TO THE TRISTEZA VIRUS*

H.M. ZUBRZYCKI, A. DIAMANTE DE ZUBRZYCKI
Estación Experimental Agropecuaria INTA,
Corrientes,
Argentina

Abstract

INDUCTION OF MUTATIONS IN CITRUS TO PRODUCE RESISTANCE TO THE TRISTEZA VIRUS.

The aim of this research is to select and isolate induced mutants of citrus that show resistance to the Tristeza virus. By means of ionizing radiations and chemical mutagens treatments were carrid out on citrus buds and citrus seeds. As the Tristeza virus produces a drastic decrease in growth in susceptible combinations as well as in young plants, this effect has been used to select and isolate resistant mutants. Basically the methods used are as follows: (a) In order to obtain a resistant cultivar, V_1 irradiated buds from a susceptible variety are grafted singly on to rootstocks resistant to the virus. In this manner V_2 buds are obtained which are then grafted on to a susceptible rootstock where differences of growth under epidemic conditions are established. As soon as a resistant plant is detected, the V_3 buds are grafted on to a susceptible rootstock to recheck the reaction and to evaluate its commercial value. (b) In order to obtain a resistant rootstock, M_1 seedlings from irradiated seed of a susceptible rootstock are grafted with susceptible variety buds. When a healthy plant is identified the grafted tree top is cut off, thereby inducing rootstock budding and seed production. The M_2 seedlings are later rechecked. During growth the (a) and (b) plants are submitted to natural infection in the presence of the transmitter aphids of the Tristeza virus. It was observed that the growth of the Valencia Late sprouts grafted on Sour Orange seedlings treated with chemical mutagens eventually showed a vegetative growth; this would prove a certain variability in the first stages of growth.

* Research supported by the IAEA under Research Contract No. 1731. The abstract only is published here, since the full paper is not available for publication.

111

PRODUCTION OF MUTANTS BY IRRADIATION OF IN VITRO-CULTURED TISSUES OF COCONUT AND BANANA AND THEIR MASS PROPAGATION BY THE TISSUE CULTURE TECHNIQUE*

E.V. DE GUZMAN, A.G. DEL ROSARIO,
P.C. PAGCALIWAGAN
Department of Horticulture,
College of Agriculture,
University of the Philippines
    at Los Baños,
Laguna, Philippines

Abstract

PRODUCTION OF MUTANTS BY IRRADIATION OF IN VITRO-CULTURED TISSUES OF COCONUT AND BANANA AND THEIR MASS PROPAGATION BY THE TISSUE CULTURE TECHNIQUE.

Regeneration of buds/shoots as well as plantlets was induced from banana shoot tip explants cultured in highly modified Murashige and Skoog's medium supplemented with coconut water and benzyladenine. Initially shoot regeneration was sparse, but on further subculture became profuse. Gamma irradiation at low dosage (1.0 kR) was stimulating to explant growth and bud formation with the two types of explants used. With Bungulan stimulation was observed even at 2.5 kR. Several morphological aberrations were exhibited by shoots of 'irradiated' in vitro plants growing in potted soil. A highly and continuously proliferating tissue strain has been isolated from a subculture which was ultimately derived from an irradiated explant. Its continued proliferation is dependent on an external supply of coconut water and benzyladenine. In vitro-produced plants have been established under field condition. The 'irradiated' plants are comparable with, and some seem to be better than, the unirradiated controls with respect to height, girth, sucker production and number of hands and fingers per bunch. Higher doses of irradiation are required to produce an adverse effect on growth of coconut embryos during the liquid culture than when growing in solid medium.

INTRODUCTION

In banana hybridization, work is hampered by problems of pollen sterility, seed sterility and poor seed germination. It is therefore hoped that the use of mutation breeding together with the tissue culture technique will accelerate progress in genetic improvement studies on the crop. The likelihood of success in inducing mutations in banana is indicated by the occurrence of spontaneous mutations affecting many

* Research supported by the IAEA under Research Contract No. 7332.

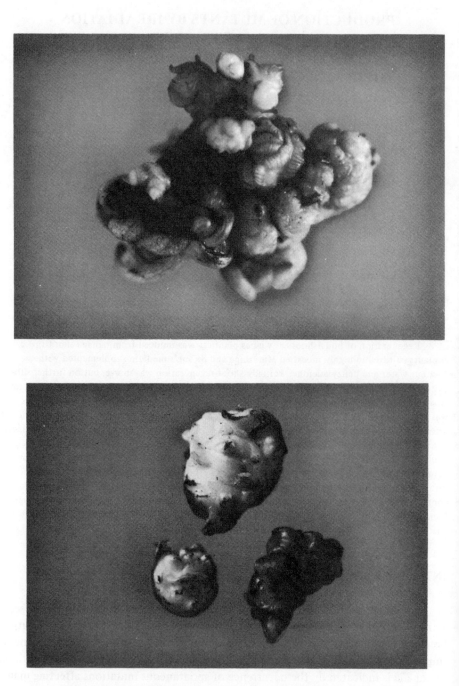

FIG.1.   Cultures of banana tissues showing clusters of regenerated buds.

TABLE I. EXTENT OF SHOOT AND BUD FORMATION FROM IRRADIATED
BANANA TISSUES IN SUBCULTURE EXPRESSED AS NUMBERS OF SHOOTS
AND BUDS PER CULTURE[a]

Irradiation dosage (kR) Variety	Irradiation dosage (kR)		
	0	1.0	2.5
Lacatan			
1st	3.1(39,15)	3.51(33,18)	1.7(27,8)
2nd	3.2(44, 15)	8.9(32, 52)	2.48(21, 15)
3rd	10.12(42, 100)	10.62(29, 88)	5.15(13, 38)
4th	20.62(94, 29)	20.18(103, 25)	4.12(9, 8)
Bungulan			
1st	2.71(31, 12)	3.5(2, 6)	2.67(9, 10)
2nd	3.89(27, 30)	2.44(9, 13)	5.53(13, 10)
3rd	2.62(40, 14)	11.94(16, 84)	16.43(30, 115)
4th	23.0(38, 31)	35.5(26, 60)	27.16(31, 101)
Saba			
1st	1.38(13, 4)	5.33(3, 8)	1.8(5, 4)
2nd	1.78(9, 5)	1.0(6, 1)	6.4(5, 15)
3rd	1.0(3, 1)	1.0(1, 1)	9.0(4, 13)
4th		1.0(1, 1)	17.25(6, 32)
D. Cavendish			
1st	2.72(25, 8)	2.14(7, 8)	—
2nd	6.33(21, 18)	2.28(7, 6)	—
3rd	6.34(35, 60)	1.25(16, 3)	—
4th	23.5(25, 42)	4.0(6, 12)	—
G. Cavendish			
1st	3.17(6, 10)	3.17(12, 5)	0.9(10, 1)
2nd	1.2(5, 2)	4.78(14, 25)	1.0(3, 1)
3rd	8.07(14, 28)	4.37(24, 30)	1.0(2, 1)
4th	15.0(2, 29)	7.37(20, 13)	1.0(1, 1)

[a] Figures in parentheses represent the number of cultures observed, and the maximum number of shoots and buds per culture respectively.

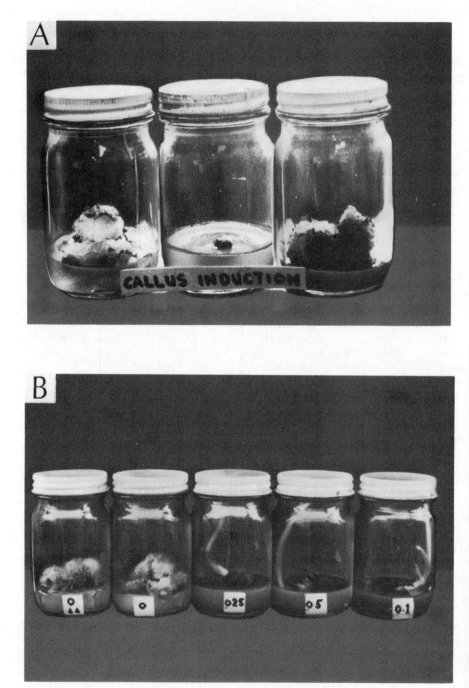

FIG.2. A. Callus induction using banana shoot tip explants Middle: initial size of explant.
B. Callus growth of banana shoot tip explants in media containing varying levels (%) of carbon.

TABLE II. GROWTH OF BANANA STEM EXPLANTS IN LINSMAIER AND
SKOOG'S MEDIUM WITH 0.1 ppm 2,4–D+1.0 ppm KINETIN AND DIFFERENT
LEVELS OF CARBON

Carbon level (%)	Trial	Degree of callus formation (%)				Shoot formation (%)
		None	Slight	Moderate	Profuse	
0 (dipped in 1% AA)	a	0	42	35	21	–
	b	9	63	18	9	–
0	a	8.3	8.3	58	25	
	b	0	60	20	29	
0.025	a	100	0	0	0	21
	b	0	0	75	25	–
0.05	a	100	0	0	0	17
	b	0	83	16	0	66
0.1	a	100	0	0	0	12
	b	75	25	0	0	16
0.2	a	–	–	–	–	
	b	100	0	0	0	50

plant characters, some of which are believed to be responsible for the formation
of new varieties. As far as tissue culture is concerned, there are previous reports
to show that banana tissues are amenable to culture in vitro.

MATERIALS AND METHODS

For initial cultures, explants were obtained from suckers. These were
prepared for irradiation and inoculation following procedures previously described.
For subcultures the explants were buds or shoot tips derived from previous
cultures. As far as possible these were used singly, but in cases where they are
tightly grouped together (Fig.1) an explant may include several buds. For irradiation, the explants were established in test tubes containing coconut water
supplemented medium to select for uncontaminated cultures. They were exposed

TABLE III. GROWTH RESPONSE OF BANANA SHOOT TIP TISSUES OF DIFFERENT VARIETIES IN A SHOOT-REGENERATING MEDIUM[c]
Observations taken after the second transfer

Variety \ Irradiation dosage (kR)	Average of buds and shoot/culture[a]			Percentage basal enlargement[a]			Percentage rooting[a]			Percentage survival[b]		
	0	1.0	2.5	0	1.0	2.5	0	1.0	2.5	0	1.0	2.5
Lacatan	13.22(60)[c]	7.04(42)	7.36(49)	69.53	80.14	50.56	59.96	89.28	64.01	53.3	63.27	69.74
Bungulan	7.07(34)	9.73(42)	9.95(40)	56.62	39.63	37.16	68.86	53.78	43.11	57.91	59.63	67.24
Saba	3.49(48)	1.94(36)	1.98(31)	83.89	51.78	80.1	48.09	27.33	20.14	70.53	56.58	73.21
D. Cavendish	9.97(41)	10.6(33)	8.61(30)	57.94	50.77	27.68	80.14	84.40	46.28	56.91	50.15	44.04
G. Cavendish	5.57(42)	7.5(47)	3.33(33)	35.37	55.36	43.79	90.29	83.21	86.77	71.23	72.62	69.68

a   Based on 3 to 6 trials.
b   Based on 2 to 4 trials.
c   Figures in parentheses are the total number of cultures observed.

TABLE IV. EFFECT OF IRRADIATION ON THE AVERAGE FRESH WEIGHT (g) OF BANANA SHOOT TIP TISSUES OF DIFFERENT VARIETIES IN A SHOOT-REGENERATING MEDIUM

Observations taken after the second transfer culture

Trial Variety	Irradiation dosage (kR)														
	0					1.0					2.5				
	1	2	5	6	Ave.	1	2	5	6	Ave.	1	2	5	6	Ave.
Lacatan	12.45	8.2	–	10.26	10.42	10.04	6.23	–	11.76	9.34	10.45	8.96	–	–	9.70
Bungulan	6.61	8.34	–	2.67	5.87	8.52	6.64	–	–	7.58	2.62	4.68	–	–	3.65
Saba	3.91	5.19	–	–	4.55	5.88	6.87	–	–	6.38	4.12	5.54	–	–	4.83
D. Cavendish	4.78	10.19	11.97	–	8.98	7.95	8.66	–	6.4	7.67	5.31	1.45	7.91	7.48	5.54
G. Cavendish	8.10	10.52	6.58	–	8.4	5.67	10.66	8.29	–	8.2	4.3	9.08	–	–	6.69

TABLE V. GROWTH RESPONSE TO IRRADIATION OF SHOOT TIPS DERIVED FROM IN VITRO REGENERATED PLANTLETS
Observations taken eight weeks after irradiation[a]

Irradiation dosage (kR) Variety/trials	Percentage survival				Number of buds and shoots/culture				Number of leaves of mother plant				Length of shoot (mm)			
	0	1.0	2.5	5.0	0	1.0	2.5	5.0	0	1.0	2.5	5.0	0	1.0	2.5	5.0
Lacatan																
I	100	100	100	83.33	1.17(12)	1(42)	2.75(16)	1(12)	3.67	3.5	3.25	0.25	44.64	45.0	38.67	0.5
II	100	100	100	100	2.54(11)	1.0(6)	2.15(13)	0.4 (5)	2.37	3.67	4.36	0	38.37	59.33	27.91	14.5
Bungulan																
I	90.91	100	91.67	57.14	1.27(11)	1.07(15)	1.7(10)	0.14(14)	2.6	3.37	3.17	0.125	42.92	44.0	48.78	7.5
II	100	100	100	66.67	1.08(13)	1.42(12)	3.33(9)	1(1)	3.31	3.5	5.22	2.0	56.08	32.86	30.8	7.5
D. Cavendish																
I	100	100	100	80	1.17(6)	1(9	1(6)	0(10	2.0	4.22	4.83	0.2	27.0	36.67	45.2	0
II	100	100	100	33.33	3.2(5)	1(10)	1(4)	1(1)	3.6	3.5	2.5	1.0	35.0	46.6	39.25	5.0
G. Cavendish																
I	71.43	100	75	85.71	1.33(6)	1(10)	1.12(10)	0	3.2	3.3	5.08	0.5	32.67	44.56	40.72	0
II	87.5	100	100	66.67	1(7)	1.33(3)	9(5)	0.33(3)	2.71	2.33	3.0	0	39.28	31.0	27	0

[a] Figures in parentheses are the number of cultures observed.

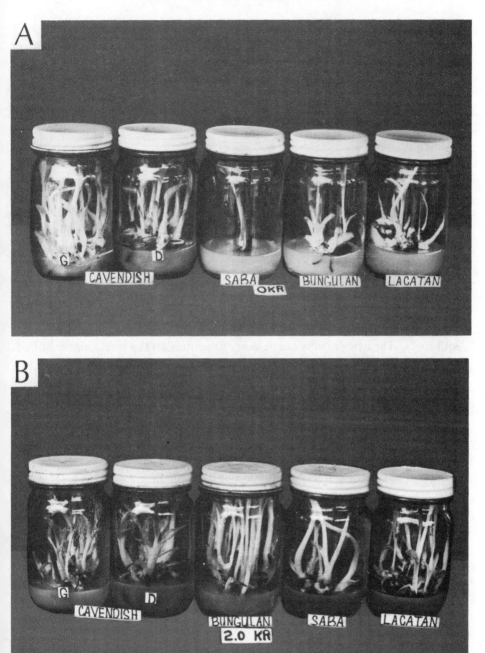

FIG.3. A. Comparative growth response in culture of banana shoot tips of different varieties.
B. Comparative growth response in culture of irradiated banana tissues of different
varieties.

to irradiation in this condition. The explants were transferred to a regeneration media the day following irradiation. For most of the experiments the medium consisted of a highly modified Murashige and Skoog's medium containing 4% dextrose, 15% coconut water and 5 ppm benzyladenine (5 BA).

RESULTS

Development of culture technique

A highly modified Murashige and Skoog's medium including coconut water (15%) and benzyladenine (5 ppm) was found satisfactory for the regeneration of new shoots. Similarly satisfactory results might be obtained with the use of the mineral formulation of Vacin and Went. The importance of cytokinins as supplement for the purpose of bud formation was previously observed by Ma and Shii (1972) in their culture of decapitated banana shoot tip sections.

In culture the explant greatly increased in size. The original shoot continued to develop and in addition several new ones were formed. After two transfers the original explant would have formed several buds, some developing into shoots with roots. The new growths can be easily subcultured. The total number of buds/shoots per culture is rather low during the initial and transfer culture periods, but on continued subculture some cultures were observed to produce a great number of shoots and buds (Table I). With the increasing tendency for bud formation and the increased amount of tissue available for subculture, the number of explants which could be derived from a single sucker is innumerable.

A preliminary attempt to induce callus formation gave very low survival and poor growth of the explant due to intense discoloration of the tissues. Callus formation was optimal at a concentration of 0.1 ppm using 2,4–D. A greater percentage of callus formation was achieved in medium without carbon and with tissues dipped in ascorbic acid before inoculation (Table II, Fig.2). In media with 0.025% carbon both callus induction and shoot formation occurred. At higher levels of carbon there was only shoot formation. Callus arises from the young scale leaves of the explant. The resulting growth was very friable and has not been successfully subcultured.

Effects of irradiation on growth

Initial trials showed that 20.0 kR was completely lethal to the explants obtained from suckers, and that the highest dosage at which viable cultures were obtained was 10 kR. On the other hand, slight growth stimulations may be effected by low dosage such as 1.0 kR and in a few cases even at 2.5 kR. The stimulating effect can be seen in the increase in weight of the cultures and in the

FIG.4. *Plant regenerated from an irradiated banana tissue showing aborted emergence of new leaves and abnormal leaf bases.*

extent of shoot formation. Irradiation-induced growth promotions were observed in cultures of explants obtained from both suckers and plantlets regenerated in vitro (Tables I, III—V, Fig.3).

Of the different varieties studied, some differences may be noted regarding their sensitivity and responses to irradiation at low dosages. For example, for the variety Bungulan slight stimulations of shoot regeneration can be observed up to 2.5 kR. With Saba there was no drop in fresh weight at 2.5 kR. Another difference among the varieties concerns their growth responses under in vitro culture. The Saba variety differs from all the other varieties in its lesser ability for shoot formation. Growth in Saba is more of the enlargement of the explant. Shoot

FIG.5. Plant regenerated from irradiated banana tissue showing bulb-like young shoot emerging through the base of non-overlapping leaf sheaths.

formation was much less compared with that in other varieties. This type of response is true with both unirradiated and irradiated Saba tissues. Three different Saba strains gave a similar type of response.

The use of shoot tips from plantlets regenerated in vitro was resorted to because of the desire to minimize variations in the explants. Uniformity of the plantlets is easier to control than that of the suckers, thus the use of the former as a source of explants is expected to give more consistent results. In addition, there are more plantlets available than suckers. In the meantime only regenerated

FIG.6. Corrugated leaf lamina cupping upwards from a plant regenerated from an irradiated banana tissue.

shoots from unirradiated controls are being used. Eventually those that come from irradiated explants will be used, hence this in effect will be giving a second dose. Preliminary observations are given in Table V. Growth was based on the extent of shoot formation, number of leaves produced by the mother shoot, length of the mother shoot and survival of culture. There was greater survival of cultures among explants obtained from plantlets. This is perhaps because with the plantlets the shoot tips are used intact and not cut up into quarters as is the case for the suckers. Vigorous cultures, some showing more shoot formation or more leaf production

FIG.7. Portion of a plant regenerated from an irradiated banana tisseu showing bilateral arrangement of leaf sheaths.

and growth, were observed up to 2.5 kR with the different varieties. Irradiation with 5.0 kR is definitely inhibitory in all types of growth response for all varieties. Except for the stunting there are no marked morphological differences.

Morphological observations

In conformity with the usual leaf character of the crop, the plantlets possess leaves divided into the lamina and leaf sheath portions. The stalk of the plantlet is a pseudostem composed of overlapping leaf sheaths. With the plantlets derived from irradiated tissues few unusual forms of the plant have been noted. Distinct morphological differences involving the leaves were noted after the plants were established in open pots in the greenhouse. One effect of irradiation is the inhibition of emergence of the younger leaves (Fig.4). In severe cases elongation of the stalk and formation of the lamina was very inhibited so that the growing portion of the shoot is reduced to a bulb-like structure; instead of emerging out of the top of the pseudostem the developing shoot bursts through the base (Fig.5). Figure 4 also shows a plant where the older leaves possess a flattened fringed petiolar base. In some cases the usual morphology was only slightly modified such as in the slight cupping and corrugations of the lamina (Fig.6) or in the flattened and bilateral arrangement of the upper leaf stalks (Fig.7). The fact that more abnormality occurs with younger leaves indicates that the younger leaf primordia and the apical meristem itself were more affected. The abnormal plants referred to above died either in the potted condition or after transplanting to the field. Some unusual features have been noted with the fruits, and these will be described below. Except for faint yellowing no chlorosis has been observed among the potted plants. White buds but not chlorotic shoots appeared in culture.

Isolation of a highly proliferating strain

A 'mutant' tissue strain that is characterized by the profuse proliferation of nodular bodies has been isolated from among the cultures derived from an irradi-ated explant. The initial rapidly proliferating culture was first noted among several cultures derived from an explant irradiated at 2.0 kR, kept in prolonged culture (eight months) in Murashige and Skoog's medium supplemented with benzyladenine and coconut water, and then subcultured twice in basal medium. The proliferations appear as tiny nodular bodies or as small buds. The growths appeared as water-soaked, white or green. The characteristic profuse nodulation persisted even in subcultures resulting in cultures with masses of nodules, buds and shoots. To test whether or not the growth of the tissues has become independent of an external supply of growth factors, a series of subcultures was performed using media given no, partial or full supplements. In media with both coconut water and benzyladenine, profuse proliferative growth occurred up to the last subculture (6th) under observation (Table VI). Proliferations may continue for some time in media without both supplements (Mo) or with coconut water only (Mc), but the total growth was decreased compared with the complete medium (Mc + BA) (Table VI and Fig.8). On further subculture those cultures consisting purely of buds or nodular bodies were no longer observed in the medium without coconut water and benzyladenine (Table VI). Proliferations of new bodies must have ceased and growth now was in the form of development of existing bodies to shoots. The results indicate that the mutant tissue strain retains the capacity for profuse bud proliferation for quite a long time, but the

TABLE VIa. FREQUENCY OF CULTURES SHOWING VARIOUS DEGREES OF BUD PROLIFERATION IN DIFFERENT MEDIA

Mo = Murashige and Skoog's basal medium, Mc = Mc plus coconut water 15%, McBA = Mc plus ppm benzyladenine; inoculum tissue is of a highly proliferating strain

Subculture series	Slight			Moderate			Profuse		
	Mo	Mc	McBA	Mo	Mc	McBA	Mo	Mc	McBA
1. (17/4/78)[a]	12.0(25)[b]	36.36(22)	26.08(23)	72	50	34.78	16.0	13.64	36.36
2. (16/5/78)	72.73(22)	47.83(23)	20.0(25)	13.64	43.48	36.0	13.64	9.09	40.0
3. (22/6/78)	100 (8)	86.67(15)	60.0(15)	0	13.33	40.0	0	0	0
4. (24/7/78)	100 (6)	100(8)	14.29(14)	0	0	64.28	0	0	21.43
5. (5/9/78)	90.91(11)	50(6)	0	9.09	50	100	0	0	0
6. (18/10/78)	0	0	0	0	0	33.33	0	0	66.67

[a] Dates of inoculation.
[b] Figures in parentheses are the number of cultures/treatment.

TABLE VIb. OCCURRENCE OF BUD PROLIFERATION AND PLANTLET GROWTH IN SUBCULTURES OF HIGHLY NODULATING BANANA TISSUES IN DIFFERENT MEDIA

Mo = Murashige and Skoog's basal medium, Mc = Mc plus coconut water 15%, McBA = Mc plus ppm benzyladenine

Subculture series	Percentage pure buds proliferation			Percentage mixed buds and shoots			Percentage pure shoot growth		
	Mo	Mc	McBA	Mo	Mc	McBA	Mo	Mc	McBA
1. (17/4/78)[a]	64(25)[b]	68.18(22)	68 (25)	36	4.54	32.0	0	0	0
2. (16/5/78)	25(24)	25.0(24)	40(25)	66.67	66.67	60.0	8.33	4.17	0
3. (22/6/78)	50(18)	0(18)	0(15)	50.0	83.33	100	0	16.67	0
4. (24/7/78)	0(19)	0(16)	9(18)	33.33	50.0	77.78	61.11	50.0	22.22
5. (5/9/78)	0(20)	9(19)	0(7)	55.0	31.58	14.28	45.0	68.0	85.71
6. (18/10)78)	0(14)	9(12)	9(12)	0	0	50.0	100	100	50.0

[a] Dates of inoculation.
[b] Figures in parentheses are the number of cultures/treatment.

ability is realized with an external supply of benzyladenine. The 'mutation' has not resulted in the operation of a self-sufficient cytokinin synthesizing system, hence the need for exogenous supply.

Other highly bud-proliferating cultures have been isolated as a result of continuous subculture. Their degree of proliferative growth, however, is below that of the first mutant tissue. The occurrence of these apparently different tissues was more common among cultures derived from irradiated tissues. Most of them come from the Lacatan variety.

The first highly proliferating tissue to be isolated is now being used as source for plantlet production. Some of the shoots that have already been developed have been grown in various kinds of media to determine the one most suited for plantlet development. Many plantlets have been produced from the mutant tissue; some have been planted out in open pots. So far these have not exhibited any marked morphological abnormalities nor signs of extra vigour. It has been observed, however, that whereas some cultures produced many shoots there are also those with only few. In a few cultures the nodules remained as tiny growths. In other words, the proliferated mass consists of buds with different growth potentials.

Observations on plants under field conditions

Vegetative growth, suckering, flowering and fruiting among the in vitro produced plants follow the pattern showed by normally propagated plants. The fruit harvested from a plant derived from an irradiated explant has the form, colour and taste characteristic of the variety. Approximate age at flowering is around 400 days from field planting. Unfortunately a number of typhoons passed by the area blowing down several of the plants.

No marked morphological difference could be observed among the surviving plants. Some of the 'irradiated' plants showed increased production of suckers, increased height and girth (Table VII).

The number of hands per bunch varied from four to seven. There were more 'irradiated' plants with seven hands per bunch. Similarly there were cases of a slight increase in total number of fingers in the irradiated treatment. The weight per bunch, however, was greater in the controls showing that the

(text cont. on p.137)

FIG.8. A. Culture of a highly proliferating strain of banana tissue.
  B. Effect of supplements on the continued proliferation of the highly proliferating tissue strain of banana.
Mc + 5 BA = *medium with coconut water and benzyladenine;*
Mc = *with coconut water;*
Mo = *without either supplement.*

TABLE VII. OBSERVATIONS ON IN VITRO-DEVELOPED BEARING PLANTS UNDER FIELD CONDITIONS

Variety and irradiation (kR)	Number of suckers per hill	Height from base to bunch emergence (m)	Diameter of pseudostem (cm)	Number of functional leaves	Number of hands	Number of fingers	Weight/ bunch (kg)
Lacatan							
0	6	3.06	21.3	6	6	82	32
	8	3.01	22.0	6	7	125	40.2
	6	4.03	22.0	8	6	74	24.5
	6	3.64	25.0	10	7	99	17.75
	8	3.38	21.5	5	4	42	11.0
	–	3.7	22.0	9	6	79	30.75
	–	3.45	23.0	7	6	69	18.75
	10	3.5	29.0	–	4	34	6.0
	7	–	31.5	6	6	87	–
	14	3.0			5	48	–
Average	8.125(8)	3.42(9)	24.14(9)	7.125(8)	5.7(10)	73.9(10)	22.62(8)
0.1	2	4.0	31.7	5	6	70	1
2.0	7	3.0	19		4	57	11.0
	10	3.73	25		6	65	12.5
	6				6	77	–
	6				6	83	–
	5				5	46	–
	12				6	84	–
	13				6	105	–
Average	8.57(7)	3.365(2)	22.0(2)	–	5.57(7)	73.86(7)	11.75(2)

2.5	5	3.75	31.2	6	7	125	40
3.0	–	3.51	23.0	9	6	97	18
4.0	8	4.66	21.55	–	6	53	15.8
5.0	9	3.55	21.5	7	–	–	–
1.0/2.0	8	3.4	42.5	5	7	98	–
2.0/2.0	12	4.31	41.0	6	7	100	–
2.5/1.0	11	3.85	36.0	2	7	102	–
		4.01	37.5	2	7	94	–
Average		3.93(2)	36.72(2)	2(2)	7(2)	98(2)	
Bungulan							
0	6	4.05	23	7	4	50	–
	8	4.66	21.55	–	4	49	–
	16	4.5	34.0	2	4	32	–
	10(3)	4.403(3)	26.18(3)	4.5(2)	4(3)	43.67(3)	

TABLE VIIIa. EFFECT OF IRRADIATION ON THE DEVELOPMENT OF COCONUT EMBRYOS IN VITRO; ONE WEEK AFTER TRANSFER TO SOLID MEDIUM, NINE WEEKS AFTER INITIAL INOCULATION
Observations after eight weeks in first solid

	Dose (kR)	Shoot		Root		Percentage survival
		Length (mm)	Number of leaves	Length (mm)	Percentage rooting	
Remained in irradiated medium	0	39.8(15)	3.33	42.07(14)	93.33(15)	90.00(20)
	0.5	34.20(20)	3.10	16.86(15)	75.00(20)	100.00(20)
	1.0	19.00(18)	2.33	10.46(13)	61.11(18)	59.89(19)
	3.0	3.2(5)	1.00	2.00(3)	60.00(5)	15.00(20)
Transferred to fresh unirradiated medium after irradiation	0.5	25.00(13)	2.84	15.41(12)	85.71(14)	77.77(18)
	1.0	19.26(19)	2.43	15.00(13)	68.42(19)	68.42(19)
	3.0	3.83(6)	1.00	4.00(2)	33.33(6)	15.79(19)

TABLE VIIIb. EFFECT OF IRRADIATION ON DEVELOPMENT OF COCONUT EMBRYOS IN VITRO; EMBRYOS IRRADIATED
THREE WEEKS AFTER TRANSFER TO SOLID MEDIUM

Observations after eight weeks in first solid

	Dose (kR)	Shoot			Root		Percentage survival
		Length (mm)	Number of leaves	Length (mm)	Percentage rooting		
Remained in irradiated medium	0	46(14)	4(14)	15.35(14)	100.00(14)		100.00(14)
	0.5	39.23(13)	3.76(13)	26.75(12)	92.30(13)		92.85(14)
	1.0	22.00(13)	2.70(13)	15.55(9)	70.00(13)		71.42(14)
	3.0	21.76(12)	2.72(12)	34.58(12)	92.30(13)		57.14(14)
Transferred to fresh unirradiated medium after irradiation	0.5	41.23(13)	3.07(13)	10.91(12)	92.39(13)		100.00(13)
	1.0	30.00(14)	3.50(14)	9.42(14)	100.00(14)		71.43(14)
	3.0	16.08(12)	1.91(12)	12.63(11)	91.66(12)		28.57(14)

*FIG.9.   A. A stand of in vitro-raised banana plants under field conditions.
          B. Close-up of fruits of an in vitro plant.*

TABLE IX. THE GROWTH OF COCONUT EMBRYOS AS AFFECTED BY
IRRADIATION
Embryos irradiated one week after initial inoculation[a]

Irradiation dosage (kR)	Fresh weight (g)		Percentage survival		Percentage callusing	
	Trials					
	I[b]	II[c]	I[d]	II[e]	I[d]	II[e]
0	0.465(4)	0.443(10)	75	100	33.33	46
0.5	0.513(14)	0.529(10)	87.5	100	12.5	11.76
1.0	0.674(16)	0.539(10)	46.15	87.5	23.08	6.25
2.0	0.452(17)	0.407(10)	76.92	77.8	0	0
4.0	0.646(23)	0.342(10)	77.78	68.75	27.77	0

[a] Figures in parentheses are the number of embryos.
[b] Taken after eight weeks in White's medium (one week solid and seven weeks liquid).
[c] Taken after five weeks in White's medium (one week solid and four weeks liquid).
[d] Taken after eight weeks in solid callusing medium.
[e] Taken after seven weeks in solid callusing medium.

individual fingers are smaller. Some misshapening of the immature fruits disappeared with maturity. Another unusual character noted was the form of the bunch stalk. In the 'irradiated' plant it forms a slight spiral whereas it drops off in a straight manner in the control. Figure 9 shows a group of in vitro raised plants growing in the field, and a close-up of a bunch of fruits which is ready for harvest.

Coconut tissue culture

Not much progress can be reported on coconut tissue culture because of the difficulty in development of culture techniques. Some encouraging proliferative growth with coconut embryos has been obtained in earlier experiments but for a long time this limited success could not be repeated. More recently, however, embryo proliferations have been induced consistently and the proliferated tissue can be subcultured with some success. Also recently we have succeeded in inducing floral meristems to revert to the vegetative conditions in vitro.

138 DE GUZMAN et al.

Some observations indicate that the growth and development of the coconut embryo is inhibited by irradiation except at very low dosage (Tables VIII and IX). This is especially so if the irradiation is administered at a stage of rapid development such as after the initial liquid culture. During the liquid culture when growth is mainly general enlargement, slightly higher dosage is required to produce growth inhibition. The irradiated embryos are still under observation for their callus growth response.

IMPROVEMENT OF SUGAR-CANE
THROUGH INDUCED MUTATIONS*

D. JAGATHESAN
Sugar-Cane Breeding Institute,
Coimbatore,
India

Abstract

IMPROVEMENT OF SUGAR-CANE THROUGH INDUCED MUTATIONS.

Results obtained on the use of induced mutations in sugar-cane breeding are summarized. Six commercial varieties under cultivation in India were subjected to mutagenic treatment for inducing mutations for specific characters. More than 50 mutants for various morphological characters, disease resistance and higher sugar content were obtained in these varieties. They were multiplied and studied for their stability for four to five years. Mutants of economic value include glabrous leaf sheath, non-flowering, vigorous and high-yielding mutants in Co 527, high-sugared and early maturing mutants in Co 419 and mutants for smut and disease resistance in Co 1287 and Co 740. Two mutants, one in Co 527 and the other in Co 419, have entered the All India Co-ordinated trials because of their superiority in yield and quality over the parent variety. Smut-resistant mutants of Co 1287 and Co 740 are being evaluated in large-scale trials. Tissue culture techniques have been used for propagating the mutants. Genetic variability has also been created by obtaining plants from callus culture with different chromosome numbers.

INTRODUCTION

About 60% of the sugar produced in the world is from sugar-cane, with India having the largest area under sugar-cane. Breeding work on this crop is being done in several countries including Australia, the United States of America, Cuba, Barbados, the Philippines, Indonesia and Taiwan. In India, the Sugar-Cane Breeding Institute has been the main breeding centre for over 65 years, and scores of Co canes/varieties in cultivation in several countries have been bred here. However, research on the use of induced mutations in sugar-cane breeding programmes has been limited. This is probably due to chimera formation and instability of mutants in vegetative propagation following irradiation of multi-cellular apices in buds, intra-somatic or diplontic selection and subsequent difficulties in isolating stable mutants without altering other traits.

Research on the use of induced mutations in sugar-cane was started in the Institute in 1968 and was intensified in 1975 (with the contract made available by the IAEA) with a view to improving a few of the commercial varieties under

---

* Research supported by the IAEA under Research Contract No. 1680.

139

TABLE I. SUGAR-CANE VARIETIES TREATED

Sample No.	Variety	Mutagen used	Mutations desired
1	Co 419	gamma rays (3, 5 & 7 krad)[a] X-rays (1–5 kR)	Earliness
2	Co 527	gamma rays (3 & 5 krad) X-rays (2–7 kR)	(a) Glabrous leaf sheath (b) Non-flowering (c) Early maturity
3	Co 1287	gamma rays (3, 5 & 7 krad) EMS (0.10–0.80%)	(a) Smut resistance (b) Increase in sucrose content
4	Co 740	gamma rays (3, 5 & 7 krad) EMS (0.10–0.80%)	(a) Smut resistance
5	Co 775	gamma rays (3, 5 & 7 krad)	Glabrous leaf sheath
6	Co 658	gamma rays (3, 5 & 7 krad)	Red-rot resistance

cultivation for specific characters since conventional breeding methods such as biparental mating and back-crossing are both time-consuming and difficult for adding desirable characters or deleting undesirable characters to a variety because of high polyploidy and the heterozygous nature of the crop. It takes about eight years to release a variety for commercial cultivation from crossing and raising of F_1 seedlings because of the lengthy process of selection at various stages. In addition, to obtain a genotype like the parent variety with added character(s) thousands of seedlings have to be grown without the certain possibility of producing one. Hence mutation breeding has a potential in this crop where the main stem or cane is the essential part for grower and industry. Any morphological mutant can be easily isolated and propagated through vegetative cuttings and need not pass through the gametic sieve as in cereals or fruits and vegetables.

MATERIALS AND METHODS

The list of varieties treated with the mutagens and the mutations desired in these varieties is given in Table I.

Single-budded sets were treated with mutagens, planted in small pots until germination was complete, and then transplanted to the field after 45 days. The first vegetative mutation generation (MV_1) data were recorded periodically up to harvest (12 months). The MV_2 generation was raised by planting cuttings from MV_1. The selected mutants were further multiplied and studied for three to four years to ascertain their stability. The promising ones were included in yield trials. The design was RBD with four replications and the plot size was four rows 3 m long.

Testing for disease resistance was done by artificial inoculation with pathogen and then scoring for infection at the time of maturity. This was repeated for four years and all the infected stools/clumps were discarded. Those which were free from infection in the fifth year were put on a preliminary yield trial along with the parent variety and then promoted to large-scale yield trials at different locations under the All India Co-ordinated Research Project for Sugar-cane.

RESULTS AND DISCUSSION

Fifty-two mutants have been isolated in Co 419, Co 312, Co 527, Co 453, Co 602 and Co 997. They include mutations for several morphological characters. They are striped mutants, leaf sheath mutants, drastic mutants, short plant-type mutants, non-flowering mutants and high-sugared mutants. Some information on these has already been published elsewhere. Reduction in plant height, number of internodes, leaf length and width increase or decrease in leaf angle and non-flowering character are some important changes in these mutants.

The bud that is treated with mutagens has a central region comprising 15–20 meristematic cells covered by 4–5 leaf sheaths; stalks developing from this are not uniform and hence chimeras result which are difficult to identify and maintain without reversion or segregation. In addition, the mutated cells may also be subjected to diplontic selection. When MV_1 and MV_2 generations were scored for changes or mutations, it was observed that within the clumps some canes showed changes and others were normal. The subsequent generations were raised by planting cuttings from all the canes. However, this resulted in instability of observed changes, and some of these never reappeared indicating that they were chimeras. In order to overcome this, mutants were selected on the basis of changes observed in individual stalks and carried over to the next generations on the stalk basis only. Here there was the advantage of stalks being used for propagation and of economic value. With this type of selection there was a remarkable increase in the number of mutants. The data have been summarized in Tables II and III. The increase in mutation frequency suggests that initials for stalks are already present in the dormant bud at the time of treatment with mutagens. The selection based on individual stalks was continued for three to four years until the mutants were

TABLE II. MUTATION RATE ON CLUMP BASIS

Variety	MV$_2$ generation				MV$_3$ generation			
	Control		Gamma and X-rays		Control		Gamma and X-rays	
	Population size	Mutation 100 plants	Population size	Mutation 100 plants	Population size	Mutation 100 plants	Population size	Mutation 100 plants
Co 312	1481	0.02	3410	1.90	1221	0.01	2360	2.01
Co 419	2079	0.09	3282	0.70	1973	0.10	1276	1.29
Co 997	499	—	1087	2.31	728	—	1363	1.39
Co 527	469	—	956	0.42	1428	—	840	0.41

TABLE III. MUTATION RATE ON STALK BASIS

Variety	MV$_1$ Mutation rate per 100 stalks	MV$_2$ Mutation rate per 100 stalks	MV$_3$ Mutation rate per 100 stalks
Co 312	4.52	75.82	76.08
Co 419	5.06	67.95	71.00
Co 527	3.97	85.05	88.96
Co 997	1.93	82.69	85.61
Co 775	2.25	62.09	71.29

TABLE IV. Co 527 MUTANT TRIAL 1974—75
Mean of four replications

Genotype	Brix	Sucrose	Number of millable canes per row	Weight of canes per row (kg)	Calculated C.C.S./acre (ton)[a]
Co 527	20.32	18.44	67.3	37.57	3.17
Mutant$_1$	19.42	17.73	51.0	30.53	2.28
Mutant$_2$	20.40	18.29	50.0	33.60	2.57
Mutant$_3$	19.79	18.09	56.8	30.09	2.29
Mutant$_4$	19.39	18.33	62.8	40.38	3.13
Mutant$_5$	19.28	17.45	51.8	28.65	2.10

[a] 1 acre = 4.047 × 10^3 m^2.
1 ton (long) = 1.016 × 10^3 kg.

stable or had not reverted. The 52 mutants in this mutation garden are stable and represent selections made on a cane basis.

Glabrous leaf sheath and non-flowering mutants

Glabrous leaf sheath mutants in Co 527 were isolated in 1968 in X-ray and gamma-ray treatments. They are of value when harvesting is by hand. There was segregation in subsequent generations and sometimes reversions, also indicating the

TABLE V. RESULTS OF THE TRIAL WITH MUTANTS (1975–76)

Sample No.	Variety	Mutant	Mean of two replications		
			Number of millable canes	Yield 20 ft row[a] (kg)	C.C.S./20 ft row[a] (kg)
1	Co 527	Control	61.50	31.65	4.086
2		M-8	60.50	29.15	3.498
3		M-10	65.50	30.90	3.638
4		M-11	59.50	27.35	3.756
5		M-12-1	40.50	17.85	2.184
6		M-12-2	65.50	26.65	3.347
7		M-12-3	72.00	27.65	3.378
8		M-12-4	66.00	27.85	3.582
9		M-12-5	39.50	13.55	1.209
10		M-13	78.00	37.15	5.144

[a] 1 ft = 3.048 × 10^{-1} m.

TABLE VI. ECONOMIC CHARACTERS OF Co 527 AND VIGOROUS
MUTANT M-10 (1977–78)

Sample No.	Character	Co 527	M-10	CD
1	Shoot population/20 ft row[a]	91.50	117.25	17.13
2	Brix	20.38	20.56	
3	Sucrose in juice (%)	17.90	17.73	
4	Purity (%)	87.83	86.24	
5	No. of millable canes/20 ft row[a]	48	103	24.38
6	Cane weight/20 ft row (kg)[a]	24.20	55.60	

[a] 1 ft = 3.048×10^{-1} m.

chimeral nature of the changed character. Here selection on the basis of
individual stalks had to be done for five years until stability could be obtained
in 1973. They were then tested for their yielding ability; data are given in
Tables IV and V.

It can be seen that although most of the mutants yield much less than parent
varieties, some of them are equal and a few are better than parent varieties
in yield. Mutant-13 of the Co 527 not only yields better, but does not flower also
indicating that mutation for two economically important characteristics has taken
place simultaneously to the advantage of the plant breeder.

Mutants of economic value

Vigorous mutant in Co 527

It was observed among the MV_3 generations that one particular mutant was
taller, growing vigorously and flowering earlier. This mutant was separated out
and studied. It was segregated for three years by vegetative propagation and
both vigorous and normal plants were observed. The vigorous plants were
repeatedly selected for three years after which the mutant became stable. It was
multiplied and a small experiment was laid out with two dates of planting
(28 Feb. and 3 April 1976) to study the rate of growth and flowering. The data
are given in Tables VI and VII.

The experiment was repeated at a different location in 1977. The data are
given in Table VIII. It is clear from this that there is an increase in the growth rate

TABLE VII. PERIODICAL HEIGHT MEASUREMENTS OF Co 527 AND MUTANT M-10

Mean of 10 canes per row from two 90-cm rows[a]

Sample No.	Number of days	Co 527 control (cm)	Co 527 M-10 (cm)
1	15 days (from the date of planting)	5.6	8.7
2	30 days	8.7	10.9
3	45 days	14.8	16.8
4	60 days	20.4	30.3
5	75 days	22.1	33.9
6	90 days	25.3	43.0
7	105 days	26.4	47.4
8	120 days	31.1	54.6
9	135 days	40.5	60.5
10	150 days	57.9	76.9
11	165 days	62.2	92.6
12	180 days	79.5	106.9
13	195 days	94.4	128.6
14	210 days	99.7	134.2
15	225 days	114.0	149.6
16	240 days	133.0	175.0
17	255 days	142.2	241.2
18	270 days	165.8	292.2
19	285 days	203.5	317.1
20	300 days	208.6	317.1
21	315 days	212.0	317.1
22	330 days	216.3	317.1
23	345 days	218.6	317.1
24	360 days	219.6	317.1

[a] Brix Co 527 Control 20.38   Sucrose Co 527 Control 17.90
Brix Co 527 M-10 20.56   Sucrose Co 527 M-10 17.73

Purity Co 527 Control 87.83
Purity Co 527 M-10 86.24
Weight Co 527 Control 24.2 kg per row
Weight Co 527 M-10 55.6 kg per row
Co 527 Control 48 canes per row
90 cm Co 527 M-10 103 canes per row.

already from the 15th day of planting in the mutant, but there was a sudden spurt in growth rate on 250th day or so compared with the parent variety. The mutant also stops growth earlier than the parent variety by 50 days. A cytological study of this mutant showed the absence of 2–3 chromosomes. The parent variety has 117 or 118 somatic chromosomes whereas the mutant has only 115 or 116 chromosomes. It shows that there is chromosome redundancy and a few chromosomes can be eliminated by mutagenic treatments; fortunately in the present case undesirable genes were located in these chromosomes. This mutant has also in addition a glabrous leaf sheath character which is of value where commercial harvesting is done by hand. However, because of the large chromosome number and similarity the missing chromosomes cannot be identified. In order to find out if this faster growth rate was due to increased cell division or cell volume, the buds of this mutant were germinated and longitudinally sectioned at intervals of 5, 10, 15, 20, 25 and 30 days. It is clear the cell population has increased through the increase in cell division whereas the cell dimensions were not different. From this it can be inferred that some of the genes responsible for growth inhibition have been deleted. It is evident that the mutant is stable, grows faster and yields higher than the parent variety with the same sucrose content and purity. This mutant has gone into the All India Co-ordinated trials as an entry in 1979–80.

High-sugared mutant

India is perhaps one of the few sugar-cane growing countries where payment is made to the cultivator on the basis of weight of the cane and not on the percentage of sucrose. However, in the recent years this policy has been changing and several sugar mills not only encourage cultivators to grow high-sugared varieties but also pay an extra premium for high-sucrose varieties to compensate for the loss of income to the farmer since high-sugared varieties usually yield less in terms of cane weight. The mutant FI-2, isolated in the variety Co 419, is a very high-sugared early maturing variety which also does not flower. The data on this mutant are given in Table IX. This mutant has not only high sugar in the 8th month (16.4%) but it goes on increasing until the 11th month. This means that this mutant is suitable to the cultivator for the manufacture of sugar even at the 8th or 9th month whereas the parent variety has hardly 12% sucrose at the 8th month. This mutant has also been stable and consistent in giving high yields and hence has also been included in the All India Co-ordinated trials for the year 1979–80.

Disease-resistant mutants

Smut disease caused by *Ustilago scitaminea* Syd. is one of the major diseases of sugar-cane in all sugar-cane growing countries except Cuba and Australia. It causes as much as 50% loss in yield and a much higher loss in commercial cane

TABLE VIII. YIELD OF MUTANTS (1978–79)
Mean of four replications

Sample No.	Variety	10th month			11th month			No. of canes 20 ft row[a]	Wt. in kg/20 ft row[a]
		Brix	Sucrose	Purity	Brix	Sucrose	Purity		
1	Co 419 Control	19.14	17.23	88.68	19.58	17.28	88.30	75	83.84
2	Co 419 M-19	20.20	18.20	19.11	18.89	16.64	88.10	62	65.94
3	Co 419 M-21	19.65	17.25	87.80	18.64	16.79	89.83	66	67.21
4	Co 419 M-26	19.62	17.42	88.70	19.20	17.20	89.56	51	56.45
5	Co 419 M-28	19.37	68.76	86.77	18.35	15.80	86.36	67	70.33
6	Co 419 M-29	19.10	17.11	89.57	19.93	17.80	89.38	57	62.35
7	Co 419 M-30	19.35	17.39	87.38	20.35	18.49	90.79	60	72.29
8	Co 419 M-31	19.07	17.17	90.03	18.99	17.08	89.78	63	68.48
9	Co 527 Control	18.62	16.84	90.44	19.05	17.19	90.20	74	44.30
10	Co 527 M-10	20.55	18.31	87.53	19.49	17.66	88.57	92	67.56
11	Co 527 M-12	17.57	15.68	89.14	18.52	16.54	89.25	69	37.60
12	F-I-2	21.00	19.40	92.38	22.23	20.46	92.02	58	62.31
13	Co 527 M-13	–	–	–	–	–	–	78	–

[a] 1 ft = 3.048 × 10^{-1} m.

TABLE IX. YIELD OF MUTANT FI-2 (1978−79)

Sample No.	Variety		10th month			11th month			No. of canes 20 ft row[a]	Wt in kg/20 ft row[a]
		Brix	Sucrose	Purity	Brix	Sucrose	Purity			
1	Co 419 Control	19.42	17.23	88.68	19.58	17.28	88.80	75	83.84
2	Co 527	18.62	16.84	90.44	19.05	17.19	90.20	74	44.30
3	FI-2	21.00	19.40	92.88	22.23	20.46	92.02	58	62.31

[a] 1 ft = 3.048 × 10⁻¹ m.

TABLE X.   MUTATIONS FOR SMUT RESISTANCE

Variety		Number of plants scored	Number of plants free from disease
Co 740	1st year	1443	183
	2nd year	1951	351
	3rd year	2237	399
	4th year	1857	433

sugar when the disease occurs in an epidemic form. Many developing countries such as Sri Lanka, Malaysia and some of the African countries are unable to cultivate a sugar-cane variety resistant to smut disease. Though this disease can be controlled by clean cultivation, use of disease-free materials and heat therapy the best control will be a resistant variety. Although sources of resistance to this disease are available from some other commercial varieties, and related wild and cultivated species, transfer of this character to cultivate successfully a variety is beset with difficulties because of the high polyploidy heterozygosity functioning of 2 n as well as n gametes in crosses and chromosomal mosaicism. Therefore, even if hybridization is carried out, the chances of obtaining a recombinant with some characteristics of the parent variety plus disease resistance are extremely remote even if millions of F_1 or BC_1 or BC_2 seedlings are raised at enormous cost. Mutations breeding has therefore been taken up to induce mutations for smut resistance in commercial varieties Co 1287 and Co 740. Co 740 is the wonder cane of India because of its high yield and high-sugar content recovery and ratooning ability in Maharashtra state, India. Co 1287 is also a high-yielding variety, but could not be released because of its slightly low sugar content and high susceptibility to smut disease. Many mutants have been obtained for smut resistance in this variety. They have been free from smut infection for the last five years and have been entered in large-scale trial in 1979–80 to evaluate their yielding ability before entry in the All India Co-ordinated trials. In Co 740, 433 smut-free mutants have been obtained and they are in the final year of testing. The data for the past five years are summarized in Table X.

In an annual crop such as sugar-cane it would first be desirable to induce as many mutants as possible for a particular specific characteristic like disease resistance and then to select from them for characteristics including yield, sugar content, purity, since all the mutants may not have the desirable characters and disease resistance together in one genotype. Selection will therefore have to be made among the mutants for commercial acceptability. The fact that a large number of smut·resistant mutants could be induced in Co 1287 and Co 740 shows that a wide base would be ultimately available for selection. In all these cases the basic genotype is almost the same with a change in one or two characters.

Other mutants

In the variety Co 419 several mutants for early and/or late germination, better early shoot population, a higher internode number and a higher number of millable canes have been isolated. All these are stable mutants and are on trial from 1979–80. These mutants, if the yield is satisfactory, will have a number of advantages. For example, the early germinating mutants (Table XI) would complete the germination phase earlier, avoiding repeated weekly irrigations immediately after planting and thus saving irrigation water wherever the crop is grown under irrigated conditions. On the other hand, the late germinating ones could also be useful in that water could be saved by giving them life irrigation, and even planting them under water-stress conditions and irrigating them later.

This study very clearly shows that it is possible to obtain an enormous amount of variability in commercial varieties in cultivation. The mutation breeder should study this variability intensively and select the mutants firstly for their stability and secondly for their use either directly as a variety with added desirable characters or as genetic stock in cross-breeding programmes.

Tissue culture

Tissue culture work was initiated in October 1978. Techniques for meristem culture and callus culture were standardized. The medium utilized is that of Murashige and Skoog with certain modifications to suit our purpose. For meristem culture the growing point (about 0.5 mm) was excised and planted in the solid medium without 2.4-D. The shoot apex starts growing within about two weeks without forming any callus. The plants obtained through meristem cultures are sugar-cane mosaic virus-free.

For callus cultures, developing leaves and young inflorescence are utilized. The callus starts developing after two to three weeks of inoculation. It will attain sufficient size for sub-culture within a month. For differentiation the callus is transferred to modified MS medium without 2.4-D. Within three to four weeks shoots start developing. The plantlets are formed in clumps at first; when the plantlets produce three to four leaves they are separated individually and transferred to Whites' medium for rooting. When sufficient roots are formed,

TABLE XI. GERMINATION IN MUTANTS
Mean of four replications

Sample No.	Variety		10th day	15th day	20th day	25th day	30th day	35th day	40th day
1	Co 419	Control	1.50	6.50	12.75	16.00	16.00	16.25	16.25
2	Co 419	M-13	4.75	7.50	10.25	11.00	11.75	12.50	13.00
3	Co 419	M-22	0.50	2.00	9.00	13.50	15.25	16.50	17.50
4	Co 419	M-25	5.50	11.00	16.00	18.00	19.00	20.25	20.25
5	Co 419	M-30	0.25	4.25	6.00	7.25	8.75	9.25	9.25
6	Co 419	M-38	5.50	15.75	20.00	22.25	22.75	23.00	23.00
	CD at 5%		3.99	2.32	3.98	3.39	3.61	3.65	2.95

they are transferred to soil and kept initially at 25°C. When they are established, plants are taken out and planted in the field.

Plantlets have been established so far from commercial Co canes i.e. Co 419, Co 740, Co 7704, interspecific hybrids involving *S. officinarum* X *S. robustum* and intergeneric hybrids. Plants obtained in variety Co 7704 are under trial to study the genetic variability created through chromosome numerical mosaicism.

MUTATION BREEDING IN SUGAR-CANE
(Saccharum sp. HYBRID)
BY GAMMA IRRADIATION
OF CUTTINGS AND TISSUE CULTURES*

S.H. SIDDIQUI, M. JAVED
Plant Genetics Division,
Atomic Energy Agricultural
    Research Centre,
Tandojam,
Pakistan

Abstract

MUTATION BREEDING IN SUGAR-CANE *(Saccharum* sp. HYBRID) BY GAMMA IRRADIATION OF CUTTINGS AND TISSUE CULTURES.
The sugar-cane variety Co 547, which is highly susceptible to smut disease (*Ustilago scitaminea* Syd.) and is also late maturing, was exposed to different doses of gamma radiation to study its radiosensitivity and to induce smut disease resistance and early maturity. The radiosensitivity of the variety showed that an optimum dose (LD_{50}) was 2.0 kR and the working dose range was found to be $1.5 - 3.0$ kR, whereas doses higher than 4.0 kR drastically affected the growth and germination. A broad spectrum of variability in reaction to disease resistance was observed after radiation exposure. This facilitated the isolation of disease-resistant mutants. Twenty-three mutants showing varied reaction to smut under field infection conditions were tested for two years by artificial inoculation using the dip method. From these studies 15 stable mutants were isolated. Of 15 mutants, seven showed promising performance in cane yield and sucrose contents. The tissue culture technique was used to determine the potential of different commercial clones for callusing. In test explants callusing was achieved readily and proliferation of callus was fairly good in all the clones.

With the increase in population, the demand for food is also increasing. To cope with the problem high-yielding varieties of crop plants have been evolved but stability in yield has declined because of the susceptibility of plants to diseases and pests. At times epidemics have caused failure and even wiped out susceptible high-yielding varieties, which has necessitated the replacement of susceptible varieties by resistant ones. Breeding disease-resistant varieties of sugar-cane has received considerable attention, and has been instrumental in controlling disease and producing increases in cane yield and sugar recoveries.

Improvement in sugar-cane breeding is handicapped in Pakistan because of intricate flowering behaviour. Therefore, induced somatic mutation breeding

---

* Research supported by the IAEA under Research Contract No. 1891.

155

holds promise for effective improvement of the crop. Induction of somatic mutation by mutagenesis is an important method in improving the specific characteristics of vegetatively propagated sugar-cane without altering most of the desired agronomic characteristics of commercial clones.

Tissue and cell culture of sugar-cane provide techniques for easy handling of mutants without chimera and sectors, and is regarded as an important breeding tool for asexually propagated plants. Tissue culture has potential in manipulating the plant system at the unicellular level for use in asexual plant improvement, and techniques have been developed for culturing sugar-cane tissue for the improvement of sugar-cane cultivars [1 − 5].

The present studies were carried out to ascertain the optimum working radiation dose range for induction of mutations for resistance to smut disease (*Ustilago scitaminea* Syd.) coupled with early maturity and high sugar recovery in clone Co 547. Tissue culture studies were carried out to determine the proliferation of callus and to ascertain the procedures for regeneration of sugar-cane callus with a view to using its potential to support and accelerate the mutation breeding programme.

RADIOSENSITIVITY STUDIES

The studies were carried out because of diverse responses of biological materials to ionizing radiation and to find out the optimum working dose range and tolerance to radiation. Smut disease-free single-eyed seed pieces of clone Co 547 (highly susceptible to smut disease) were treated with aglol fungicide before irradiation and were exposed to gamma radiation in 1.0, 2.0, 4.0, 6.0, 7.5 and 10.0 kR doses. Irradiated and unirradiated seed pieces were planted in an RCS design with four replications. In the MV_1 generation measurement of plant height and survival rate was made at 52 days after planting, beyond which no further germination occurred. Growth performance was obtained by multiplying the survival rate by the plant height and was used as an indicator of morphological and cytological damage. On the basis of observed growth performance, values of LD_{50} were estimated. Growth performance as a percentage of control was transformed to arcsin for analysis. Linear regression was used to establish the relationship between the radiation dose and growth performance. A working dose range was then calculated. At the time of seed cutting morphological variations were also observed.

Determination of optimum dose in sugar-cane is usually obtained by direct observation, which is time-consuming. A quick method has been suggested by Wu and co-workers [6] to predict the optimum dose at the earlier plant growth stages, to facilitate the estimation of the working dose range of radiation for mutation breeding. Conger and Stevenson [7] consider that the plantlet height

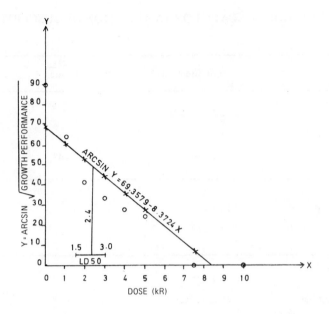

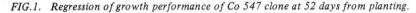

FIG.1.   Regression of growth performance of Co 547 clone at 52 days from planting.

is a valid indicator for evaluating radiation damage in spite of heterogeneity in the treatment lot.

In sugar-cane, the dosage effects are varied. Tysdal [8] reported 4.0 kR to be the lethal dose, while Panje and Parasad [9] indicated 7.2 kR had little effect with approximately 50% mortality at 14.4 kR. Clonal differences in radio-sensitivity have been reported [8 – 13].

Radiosensitivity was determined on the basis of observed growth performance by regression from analysis of variances shown in Fig.1. Linear reduction in growth in terms of plantlet height and growth performance as a result of increased dosage was highly significant. Optimum dose (LD_{50}) and dose range estimated at 52 days to induce mutation were 2.4 kR and 1.5 – 3.0 kR respectively, as indicated in Fig.1.

Analysis of variance showed that differences among radiation treatments were highly significant and negatively correlated. The growth was drastically affected by doses larger than 4.0 kR. Doses of 7.5 kR and 10.0 kR were almost lethal. The morphological changes observed were (1) multiple auxiliary buds, especially at the base of stalks; (2) chlorosis, chlorotic spots, striata, stripes and albinoism; (3) inhibition of tillering; (4) narrow leaves and thin stalks; and (5) shooting eyes.

TABLE I. NUMERICAL SCALE FOR EVALUATION OF SUSCEPTIBILITY
TO SMUT

Reaction	Stool infection (%)	Grade	Symbol
Immune	0	0	1
Highly resistant	1—5	1—2	HR
Resistant	5.1—10	3—4	R
Moderately resistant	10.1—15	5	MR
Moderately susceptible	15.1—20	6	MS
Susceptible	20.1—25	7	S
Highly susceptible	25 & above	8—9	HS

The average time required for germination was increased with increasing
dose, and germination was erratic. At low radiation doses the percentage of
viable plants was comparable with that of unirradiated but survival after
germination was drastically reduced beyond the 4.0 kR dose and survivors
showed delayed initiation of growth. The delayed growth was negatively
correlated with the radiation exposure.

RADIATION EFFECTS ON INDUCTION OF SMUT DISEASE (*Ustilago scitaminea* Syd.) RESISTANCE IN SUGAR-CANE (*Saccharum* sp.) HYBRIDS

Fifteen seed pieces from normal looking stools, i.e. free from smut whips,
were obtained from the MV_1 generation. These seed pieces were planted
randomly in rows in an area heavily infected with smut spores for the MV_2
generation, and screening of desirable mutants was carried out under natural
field infection in the MV_3 generation.

In the MV_4 and MV_5 generations disease reaction was tested by artificial
infection by dipping in a smut spore suspension of 5×10^6 spore per ml as
practised at the Hawaiian Sugar Planter's Association Experiment Station [14]
to inoculate the seed pieces. The reaction to smut disease was scored on the

TABLE II. CLASSIFICATION OF MUTANTS OF CLONE Co 547 IN SMUT TEST ON STOOL INFECTION PERCENTAGE IN MV_4

Group I Immune = 0%	Group II Highly resistant = 1–5% Resistant = 5.1–10%	Group III Moderately resistant = 10.1–15% Moderately susceptible = 15.1–20%	Group IV Susceptible = 20.1–25% Highly susceptible = 25% and above
AEARC – 2001	AEARC – 2011	AEARC – 1001	AEARC – 2015
AEARC – 2012	AEARC – 2006	AEARC – 1011	AEARC – 2013
AEARC – 3002	AEARC – 2017	AEARC – 4006	AEARC – 2019
AEARC – 4003	AEARC – 1002		AEARC – 2008
	AEARC – 4005		AEARC – 2021
	AEARC – 5002		AEARC – 5004
	AEARC – 5008		AEARC – 5010
	AEARC – 5013		Parent Co 547
	AEARC – 2005		

percentage of infected stools. The infection was recorded on the appearance of a whip.

The screening was done on the total number of smutted stools in original stand and graded according to the modified Hawaiian numerical scheme [15] listed in Table I. This method provides a disease rating which has a close correlation between natural and artificial infection by the dip method under field conditions. Smut-resistant sub-clones treated with spore suspension were randomly planted in rows in the MV_3 generation in the field for further smut testing.

The variation in disease resistance among sugar-cane clones is well known and widely exploited to produce resistant varieties. Successful use of radiation in inducing desired mutations has been reported by several sugar-cane breeders. Rao and co-workers [16] have isolated and released a red-rot-resistant mutant marked as Co 6602 from clone Co 449. Another red-rot-resistant mutant has also been reported from the Co 997 clone by the same workers. Jagathesan and co-workers [17] and Haq and co-workers [18] also reported several moderately resistant mutants for red-rot disease. Breaux [10] and Darmodjo and Seedjono [19] reported mosaic-resistant mutants and Siddiqui [20] and Jagathesan [21] have also isolated smut-resistant mutants. From a heavily flowering variety,

TABLE III. SMUT DISEASE REACTION, SUCROSE AND PURITY PERCENTAGE OF Co 547 MUTANTS DURING DIFFERENT STAGES OF SELECTION

Sample No.	Mutants	MV$_4$ Smut disease reaction	MV$_5$ Smut disease reaction	Sucrose[a] (%)	Purity[a] (%)
1.	AEARC—1002	HR	R	11.73	76.68
2.	AEARC—5002	HR	HR	10.62	71.03
3.	AEARC—4006	MR	MR	15.04	89.92
4.	AEARC—1001	MR	MR	11.95	78.02
5.	AEARC—5013	HR	R	13.23	81.11
6.	AEARC—5008	HR	HR	12.55	80.74
7.	AEARC—2012	I	I	12.55	82.51
8.	AEARC—2006	HR	HR	12.61	80.64
9.	AEARC—4005	HR	HR	9.47	70.26
10.	AEARC—2017	HR	R	11.65	77.58
11.	AEARC—2001	I	I	11.80	77.30
12.	AEARC—4003	I	I	11.73	77.56
13.	AEARC—2005	MR	HR	10.90	73.57
14.	AEARC—2001	HR	HR	—	—
15.	AEARC—3002	I	MR	12.78	79.91
16.	AEARC—1011	MR	—	—	—
17.	AEARC—2013	HS	—	—	—
18.	AEARC—2019	HS	—	—	—
19.	AEARC—2008	HS	—	—	—
20.	AEARC—2015	HS	—	—	—
21.	AEARC—2021	HS	—	—	—
22.	AEARC—5004	HS	—	—	—
23.	AEARC—5010	HS	—	—	—
24.	Co 547	HS	HS	12.37	80.37

[a] Ten-month-old cane.

one non-flowering mutant with promising yield performance was isolated by Rao [11], and Jagathesan [22] isolated a fast-growing vigorous mutant with a higher number of millable canes and earlier in maturity.

Initially 200 mutants with varied degrees of disease reaction, i.e. free from smut whip (I) to moderately susceptible, were isolated in the MV_2 generation under natural field conditions, and further screening was carried out in the MV_3 where 23 mutants were isolated with a high degree of disease-resistant reaction. These 23 mutants were tested under artificial inoculation by the dip method in the field for two years (MV_4 and MV_5 generations). The reaction to smut disease is summarized in Tables II and III.

In the MV_4 testing reaction to smut disease was varied as indicated in Table II. The mutants were classified according to their reaction based on stool infection percentage. Four mutants of group I had no infection, nine mutants of group II had 2.3% − 6.97% infection with the HR-R reaction, three mutants with 11.5% − 12.0% infection were in group III, whereas group IV consisted of susceptible clones with more than 26% infection and were discarded.

Fifteen mutants showing the I, HR, R and MR reactions were advanced in the MV_5 generation for further testing. These mutants were again infected artificially for the study of stability of disease resistance. These were also evaluated for sugar content and maturity. As summarized in Table III, all the mutants showed stability for disease reaction, and mutants AEARC-4006, AEARC-5013, AEARC-5008, AEARC-2012, AEARC-2006 and AEARC-3002 also showed an increase in sucrose content and purity percentage in the ten-month-old cane. The higher sugar in the early stage indicates the tendency for early maturity. Morphologically all these mutants were similar to the parent clone except that leaves were free from spiny hairs.

In the year 1977 − 78, 15 smut disease-resistant stable mutants were advanced in the next generation for preliminary yield trials. The analysis of variance revealed that differences were non-significant for cane yield and sugar contents among the mutants, but mutants were significantly better when compared with the parent and check varieties as summarized in Table IV. The mutants AEARC-5013, AEARC-5008, AEARC-2012 and AEARC-3002 maintained their superiority in sugar content and cane yield over the parent clone Co 547. Mutant AEARC-5008 had the highest sucrose content. In addition, mutants AEARC-1001, AEARC-2006, AEARC-4006 and AEARC-4003 showed superior performance in cane yield and sugar content.

During 1978−79, all these 15 mutants were grown at five different locations to study their yield potential under different climatic conditions. Of all the mutants, AEARC-1002, AEARC-5002, AEARC-5013, AEARC-5008, AEARC-2012, AEARC-2001 and AEARC-2011 were found with better performance in sugar content at all locations, as shown in Table V. Among the four mutants reported earlier in 1977−78, three, namely AEARC-5013,

TABLE IV. YIELD (CANE AND SUGAR) PERFORMANCE OF SMUT DISEASE-RESISTANT MUTANTS OF Co 547 AND OTHER CULTIVARS DURING 1977–78

Sample No.	Clones	Average stalks per stool	Cane yield		Sugar	
			Average wt/ stool (kg)	Average yield/ plot (kg)	Sucrose (%)	Purity (%)
1.	AEARC–1002	5.33	6.42	329.00	15.36	86.91
2.	AEARC–5002	5.00	5.98	324.52	15.08	85.04
3.	AEARC–4006	6.00	5.85	284.97	15.51	88.14
4.	AEARC–1001	5.67	5.12	284.96	16.69	88.55
5.	AEARC–5013	5.00	5.22	281.25	16.19	88.11
6.	AEARC–5008	5.00	5.54	275.97	17.31	89.57
7.	AEARC–2012	4.67	5.33	271.85	15.53	85.87
8.	AEARC–2006	6.67	5.70	267.91	16.61	89.05
9.	AEARC–4005	5.67	5.49	246.29	15.66	85.91
10.	AEARC–2017	6.33	4.98	240.82	15.81	88.22
11.	AEARC–2001	5.00	4.76	223.79	15.88	87.60
12.	AEARC–4003	5.33	4.96	210.83	16.13	85.81
13.	AEARC–2005	5.33	4.14	199.47	14.43	85.65
14.	AEARC–2011	6.00	4.40	198.70	15.99	86.69
15.	AEARC–3002	6.00	4.06	187.70	15.41	87.04
16.	Co 547 (Parent)	6.00	3.83	100.55	15.33	84.85
17.	PR 1000 (Check)	6.00	3.23	84.77	16.76	90.50
18.	BL 4 (Check)	4.67	2.13	54.26	17.07	86.34
	S.E.	0.6887	0.962	69.00	0.626	1.622
	LSD1	1.94	2.66	137.32	–	–
	LSD2	2.57	3.53	182.17	–	–

AEARC-5008 and AEARC-2012, maintained their superiority at all the five locations under study while the fourth mutant, AEARC-3002, performed better at two locations but could not exceed the parent at the remaining three places. Similarly the remaining eight mutants showed good performance at some locations but poor at the others.

TISSUE CULTURE STUDIES

In sugar-cane, cell and tissue culture techniques have been developed [1, 2, 4, 5] for the improvement of commercial clones, particularly in existing valuable clones which are generally satisfactory and have most of the superior traits, e.g. high yield, but have limited potential because of an inherent deficiency in certain characteristics: disease resistance, maturity and sugar content.

Nickell [4] reported isolation of five cellular sub-clones which were different in chlorophyll content and growth habit. These clones were cytologically analysed by Heinz and Mee [1] who found that each of them had a different range of chromosome numbers, thus for the first time demonstrating chromosomal mosaic in sugar-cane. This variability was attributed to the variation in chromosome numbers from cell to cell in the original parent and it was later established that most of the commercial varieties were chromosomal mosaics, which was confirmed by Tlaskal and Hutchinson [23]. Nickell [5] further reported sub-clones resistant to Fiji disease, eye-spot and downy mildew disease derived from plantlets generated from cell culture.

Sugar-cane varieties PR1000, NC0310, SL 4 and Co 547 were used for callusing. The explants were initially obtained from shoot apices which usually consist of the meristem dome plus several pairs of leaf primordia. The explant was thoroughly washed with tap water and older leaves were removed. The surface was sterilized with 95% ethyl alcohol and phenyl mercuric acetate (PMA 1: 400) and rinsed thoroughly with sterile distilled water. The sterile apical apices were dissected by making cuts aseptically. Sliced tissues were explanted inside sterile bottles containing MSC3 (Murashige and Skoog medium [24]) plus 10% coconut water by volume and 3ppm 2.4-D. The procedure for induction and growth of callus was as described by Heinz and Mee [1].

In the explant callusing was achieved readily and proliferation was fairly good. The callus was irradiated with a ^{60}Co source by exposing it to $1-10$ kR gamma rays. Immediately after irradiation, callus tissues were transferred to fresh MSC3 medium for further observations. The colour and friability of the callus was affected by radiation. With increasing doses the callus continued to darken with no sign of proliferation. In unirradiated callus, the differentiated plantlets were weak because of contamination and were killed in the early stages of totipotency. The contamination percentage was high and could not be controlled under existing laboratory conditions. More aseptic conditions are being tried.

TABLE V. PERFORMANCE (SUGAR CONTENT) OF MUTANTS OF Co 547 AT DIFFERENT LOCATIONS DURING 1978–79

| Sample No. | Clones | AEARC | | | | ARI | | Nawabshah | | Moro | | Mirpurkhas | | Disease reaction |
| | | Ratoon | | Plant | | Plant | | Plant | | Plant | | Plant | | |
		Sucrose (%)	Purity (%)	Sucrose (%)	Purity (%)	Sucrose (%)	Purity (%)	Sucrose (%)	Purity (%)	Sucrose (%)	Purity (%)	Sucrose (%)	Purity (%)	
1.	AEARC–1002	16.39	90.71	12.43	81.89	17.39	90.36	12.84	82.17	12.99	76.61	11.87	75.60	R
2.	AEARC–5002	18.14	93.64	14.36	88.68	18.20	91.98	12.14	81.75	12.13	76.88	13.32	78.85	HR
3.	AEARC–4006	15.67	85.98	12.95	89.29	18.15	92.77	11.41	77.53	12.51	75.69	12.16	75.53	MR
4.	AEARC–1001	16.99	92.41	12.56	84.81	17.94	90.00	11.33	73.40	12.08	73.50	–	–	MR
5.	AEARC–5013	17.89	93.45	12.89	88.05	17.74	93.62	11.44	78.78	11.92	73.42	12.11	75.69	R
6.	AEARC–5008	17.48	92.72	13.43	86.85	17.59	90.51	12.16	78.67	10.52	73.59	–	–	HR
7.	AEARC–2012	16.34	92.35	12.60	85.17	17.24	92.07	12.03	79.06	11.91	76.92	13.52	76.17	I
8.	AEARC–2006	15.34	90.83	12.86	86.84	16.04	89.46	10.76	76.74	11.85	76.08	13.31	78.99	MR
9.	AEARC–4005	17.94	91.91	11.43	81.54	17.19	91.93	11.61	79.10	11.20	74.53	12.92	79.28	HR
10.	AEARC–2017	16.79	90.77	13.45	90.95	17.57	92.55	11.96	80.80	11.05	75.24	13.56	79.23	I
11.	AEARC–2001	17.64	91.97	14.54	88.72	16.49	91.31	11.64	76.69	11.80	77.50	12.47	80.98	I
12.	AEARC–4003	16.29	92.16	12.73	85.33	17.16	89.94	11.76	79.56	11.99	74.53	–	–	MR
13.	AEARC–2005	16.19	92.56	12.70	85.25	17.72	91.84	12.29	82.05	11.57	75.94	13.50	81.32	HR
14.	AEARC–2011	18.29	93.91	11.42	85.39	16.89	90.56	12.64	83.86	12.15	75.43	13.16	80.24	I
15.	AEARC–3002	18.19	93.91	12.98	85.40	16.99	91.45	11.44	77.80	11.26	77.32	–	–	HR
16.	Co 547 (Parent)	15.59	85.11	13.03	87.11	17.84	92.00	11.61	78.65	12.57	77.23	12.49	77.75	HS

ACKNOWLEDGEMENTS

We gratefully appreciate assistance in sugar analysis from M.R. Lobhi, sugar-cane specialist, ARI, Tandojam, and B.F. Jafri of AEARC, Tandojam. We are also thankful to Raziullah Khan and A. Qayoom for assistance in field and laboratory work.

REFERENCES

[1] HEINZ, D.J., MEE, G.W.P., Plant differentiation from callus tissue of *Saccharum* species, Crop Sci. **9** (1969) 346.

[2] HEINZ, D.J., MEE, G.W.P., NICKELL, L.G., Chromosome numbers of some *Saccharum* species hybrids and their cell suspension cultures, Am. J. Bot. **56** (1969) 450.

[3] HEINZ, D.J., "Sugar-cane improvement through induced mutations using vegetative propagules and cell culture techniques", Induced Mutations in Vegetatively Propagated Plants (Proc. Panel Vienna, 1972), IAEA, Vienna (1973) 53.

[4] NICKELL, L.G., Test tube approaches to bypass sex, Hawaiian Planter's Record **58** 21 (1973) 293.

[5] NICKELL, L.G., Crop improvement in sugarcane: studies using in vitro methods, Crop Sci. **17** (1979) 717.

[6] WU, K.K., SIDDIQUI, S.H., HEINZ, D.J., LADD, S.L., Evaluation of mathematical methods for predicting optimum dose of gamma radiation in sugarcane (*Saccharum* sp.), Environ Exp. Bot. **18** (1978) 95.

[7] CONGER, A.D., STEVENSON, H.Q., A correlation of seedling height and chromosomal damage in irradiated barley seed, Radiat. Bot. **9** (1969) 1.

[8] TYSDAL, H.M., "Promising new procedures in sugarcane breeding", Proc. 9th Congr. ISSCT (1956) 618.

[9] PANJE, R.R., PARASAD, P.R.J., "The effects of ionizing radiations on sugarcane", Proc. 10th Congr. ISSCT (1959) 775.

[10] BREAUX, R.D., Radiosensitivity and selection for mosaic-resistant variants in sugarcane; Proc. Am. Soc. Sugarcane Technologist NS **4** (1975) 96.

[11] RAO, P.S., "Radiosensitivity and non-flowering mutants in sugarcane", Proc. 14th Congr. ISSCT (1972) 408.

[12] SIDDIQUI, S.H., MUJEEB, K.A., KEERIO, G.R., Gamma irradiation effects on sugarcane (*Saccharum* sp.) clone Co-547, Environ. Exp. Bot. **16** (1976) 65.

[13] URATA, R., HEINZ, D.J., "Gamma irradiation-induced mutations in sugarcane", Proc. 14th Congr. ISSCT (1972) 402.

[14] BYTHER, R.S., STEINER, G.W., "Comparison of inoculation techniques for smut disease testing in Hawaii", Proc. 15th Congr. ISSCT (1974) 280.

[15] LADD, S.L., HEINZ, D.J., Smut reaction of non-Hawaiian sugarcane clones, Sugarcane Pathologist's Newsletter **17** (1976) 6.

[16] RAO, J.T., SRINIVASAN, K.V., ALEXANDER, K.C., A red rot resistant mutant of sugarcane induced by gamma irradiation, Proc. Indian Acad. Sci., Sect. B **64** (1966) 224.

[17] JAGATHESAN, D., BALASUNDARAM, N., ALEXANDER, K.C., "Induced mutations for disease resistance in sugar-cane", Induced Mutations for Disease Resistance in Crop Plants (Proc. Research Co-ordination Meeting Novi Sad, 1973), IAEA, Vienna (1974) 151 (Short communication).

[18] HAQ, M.S., RAHMAN, M.M., MIA, M.M., AHMED,H.U., "Disease resistance of some mutants induced by gamma rays", Induced Mutations for Disease Resistance in Crop Plants (Proc. Research Co-ordination Meeting Novi Sad, 1973), IAEA, Vienna (1974) 150 (Short communication).

[19] DARMODJO, I.R., SEEDJONO, M., Induction of mosaic disease resistance in sugarcane by gamma ray irradiation, ISSCT Sugarcane Breeder's Newletter 39 (1977) 4.

[20] SIDDIQUI, S.H., Induced somatic mutation breeding through vegetative cuttings and tissue culture technique by gamma radiation, paper presented at 3rd IAEA Co-ordination meeting, Skierniewice, Poland, 1978 (unpublished).

[21] JAGATHESAN, D., Induced mutations in sugarcane, paper presented at 3rd IAEA Co-ordination meeting, Skierniewice, Poland, 1978 (unpublished).

[22] JAGATHESAN, D., RAJINI, R., A vigorous mutant in sugarcane (*Saccharum* sp.) clone Co-527, Theor. Appl. Genet. 51 6 (1978) 311.

[23] TLASKAL, J., HUTCHINSON, P.B., "An objective method for counting chromosomes in sugarcane root meristems", Proc. 15th Congr. ISSCT (1974).

[24] MURASHIGE, T., SKOOG, F., A revised medium for rapid growth and bio-assays with tobacco tissue cultures, Physiol. Plant. 15 (1962) 473.

MUTATION BREEDING OF VEGETATIVELY PROPAGATED TURF AND FORAGE BERMUDA GRASS*

G.W. BURTON, W.W. HANNA
Agricultural Research, Science
  and Education Administration,
United States Department of Agriculture
and
College of Agriculture Experiment Stations,
University of Georgia,
Agronomy Department,
Coastal Plain Station,
Tifton, Georgia

J.B. POWELL
Agricultural Research, Science
  and Education Administration,
United States Department of Agriculture,
Plant Genetics and Germoplasm Institute,
Beltsville, Maryland,
United States of America

Abstract

MUTATION BREEDING OF VEGETATIVELY PROPAGATED TURF AND FORAGE
BERMUDA GRASS.

Tifgreen, Tifway and Tifdwarf, sterile triploid (2n = 27) F_1 hybrids between *Cynodon dactylon* and *C. transvaalensis*, are widely used turf grasses bred at Tifton, Georgia. They cannot be improved by conventional breeding methods. Attempts to improve them by treating short dormant rhizome sections with EMS failed but exposing them to 7 − 9 kR of gamma radiation produced 158 mutants. These have been evaluated at Tifton, and Beltsville, Maryland, and nine that appear to be better than the parents in one or more characteristics were planted in 8 X 10 m plots in triplicate in 1977. Test results to date suggest that one or more of these will be good enough to warrant a name and release to the public. Coastcross-1 is an outstanding sterile F_1 hybrid bermuda grass that gives 35% more beef per acre but lacks winter hardiness. Since 1971, several million sprigs of Coastcross-1 have been exposed to 7 kR and have been planted and screened for winter survival at the Georgia Mountain Experiment Station. Chlorophyll-deficient mutants have appeared and one mutant slightly, but significantly, more winter hardy than Coastcross-1 has been obtained. Sprigs of this mutant named Coastcross 1-M3 are being irradiated and screened in an attempt to increase its winter hardiness.

* Co-operative investigations of the Agricultural Research, Science and Education Administration, U.S. Department of Agriculture, and the University of Georgia. Study supported in part by the U.S. Department of Energy Contract No. DE-AS09-76-EV00637. Research carried out in co-operation with the IAEA under Research Agreement No. 1240.

167

BERMUDA GRASS

Bermuda grass is one of the world's most versatile grasses. It is a highly variable genus that ranges in size from plants with pencil-sized stems that reach a height of 1.5 m to tiny fine-stemmed types that grow less than 12 cm tall. 'Common' bermuda grass, *Cynodon dactylon* ($2n = 36$), and 'African' bermuda grass, *C. transvaalensis* ($2n = 18$), are the two species best suited for turf.

Bermuda grass reproduces sexually and may be propagated by seed. However, its tough, rapidly spreading stolons and rhizomes make vegetative propagation practicable, and all improved varieties are planted in this way.

TURF BERMUDA GRASSES

Breeding bermuda grass for turf began at Tifton, Georgia, in 1942 when a very dense dwarf (from the pasture breeding research) was crossed with highly disease-resistant selections of common bermuda grass.

The superiority of one of these hybrids (Tifton 57) was proven in three years of comparison with the best selection from a number of southern golf courses in plots planted at Tifton in 1946 [1]. Characteristics sought in these grasses were dependability, good green colour throughout the growing season, frost resistance, drought tolerance, weed resistance, disease resistance and compatability with over-seeded winter grasses. Tifton 57, officially released as 'Tiflawn' in 1952, continues to be the best variety for football fields, playgrounds and other areas that receive rough treatment. It was too coarse and made too much growth for golf greens.

The next product of the turf breeding programme was 'Tiffine' (Tifton 127), a cross between Tiflawn and *C. transvaalensis* [2]. This sterile triploid ($2n = 27$) had a softer, finer texture and was better suited for golf greens than Tiflawn. It was soon replaced, however, by 'Tifgreen' (Tifton 328), an F_1 triploid hybrid between *C. transvaalensis* and a superior *C. dactylon* from a golf green at the Charlotte Country Club in North Carolina.

Tifgreen, officially released in 1956, made a better putting surface than other varieties and has been extensively planted on golf greens [3]. It has also been used to a lesser degree on fairways, tees and lawns.

The fourth improved turf variety (released in 1960) was 'Tifway' (Tifton 419) a dark green sterile triploid *(C. transvaalensis* $\times$ *C. dactylon)* hybrid with greater frost tolerance than Tifgreen [4]. Its stiffer leaves and greater pest resistance than Tifgreen make it particularly well suited for golf fairways and tees, lawns and athletic fields with moderate wear.

In 1965 'Tifdwarf', a vegetative mutant of Tifgreen, was released. Tifdwarf has finer stems, shorter internodes and smaller, softer, darker green leaves than

Tifgreen [5]. It makes a denser turf than Tifgreen and when mowed at 5 mm and properly managed makes a putting surface comparable with the best bent grasses. Although planted on lawns, it is best suited for golf greens and bowling greens.

MUTATION BREEDING

All interspecific *C. dactylon* × *C. transvaalensis* hybrids are sterile triploids (2n = 27) and shed no pollen [6]. This makes them attractive lawn grasses for people who are allergic to bermuda grass pollen. Such sterility facilitates their control, yet it imposes no serious handicap on their use because they can be easily propagated by planting sprigs. The sterility of these hybrids does, however, prevent their improvement by the common plant-breeding methods of hybridization and selection.

The success of Tifdwarf, a natural mutant of Tifgreen, suggested to us several years ago that increasing the natural mutation rate with the aid of mutagenic agents could create other useful varieties. Such mutants should retain most of the superior traits of the 'Tif' bermuda grasses while differing in such traits as plant colour, size and pest resistance. Theoretically, treatment of highly heterozygous plants, such as the 'Tif' bermuda grasses, with mutagens should create mutations that can be seen in the immediate M-1 generation.

Thus in the winter of 1969–70 research designed to create mutants in the best triploids Tifgreen, Tifway and Tifdwarf [7], was begun. Dormant sprigs (stolons and rhizomes), washed free of soil and cut into one and two node sections, were treated with the chemical mutagen EMS and gamma rays from a ^{60}Co source.

The EMS treatments failed but certain doses of gamma rays (7000 to 9000 R) produced 158 mutants (the best dose for dormant bermuda grass rhizomes appears to be 7000 R). The mutants were identified by planting irradiated sprigs that germinated in 5 cm pots in the greenhouse, transplanting them on 0.6 m centres in the field and observing them daily as they began to spread. Early morning gave the best light for identifying colour differences. The mutants were transplanted in pots in the greenhouse, increased vegetatively and planted in plots at Tifton, Georgia, and Beltsville, Maryland, where they have been evaluated for several years.

Some of the mutants have exhibited sectoring in the plots where they have been evaluated. It now appears, however, that over half of them may be considered solid mutants. Several mutants that are smaller and slower growing than Tifdwarf seem to have no value except for miniature gardens. Other mutants that seemed better than their parents Tifgreen, Tifway and Tifdwarf early in the test period now are recognized as no better if as good. New varieties must be better than those now available and satisfying this requirement is not easy.

There are still nine mutants that appear to be better than their parents in one or more important characteristics. Two of these seem to be immune to root knot nematode. Two seem able to tolerate attack from several nematode species without loss of vigour. One mutant rarely produces seed heads. These nine mutants and their three parents were increased and planted in 8 × 10 m plots in triplicate in 1977. Management variables applied to each plot beginning in 1978 include cutting height and frequency, nematicide treatments, insecticide treatments and herbicide treatments. A.W. Johnson, nematologist, and Homer Wells, plant pathologist, are assisting in the final evaluation of these mutants.

In 1979 these nine superior mutants and their three parents were planted on several golf courses located as far north as New Jersey to be evaluated for winter hardiness under golf green and fairway management. If they pass these tests, it is believed that one or more of them will be good enough to name and release as new varieties capable of replacing the excellent parents from which they came.

PASTURE BERMUDA GRASS

In 1943 'Coastal' bermuda grass was released, an F_1 hybrid that yielded nearly twice as much dry matter as the common type in good years and up to six times as much in very dry seasons [8]. It produced very little seed and had to be propagated vegetatively. Machines were developed to facilitate planting and it has now been planted on more than 4 million hectares across the southern United States of America.

In 1967 a new F_1 hybrid bermuda grass called 'Coastcross-1' [9] was released. A cross between Coastal and an introduction from Kenya, this hybrid carried much of the quality of its Kenya parent but also much of its susceptibility to winter killing. In repeated tests, it produced as much dry matter as Coastal bermuda but produced 35% more beef because it was 12% more digestible. Coastcross-1 has been very sucessful in Florida, south Texas and tropical countries such as Mexico and Cuba. It cannot be grown in most of the United States because it lacks adequate winter hardiness.

Coastcross-1 is completely sterile and cannot be made more winter hardy by conventional breeding methods. Coastal bermuda grass produces rhizomes (under ground horizontal stems) that help it to survive the freezes of short duration generally experienced in the south. The Kenya parent of Coastcross-1 has no rhizomes and Coastcross-1 rarely produces them. We have assumed, there-fore, that a dominant gene for no rhizomes from the Kenya parent is making rhizome development in Coastcross-1. If Coastcross-1 could be made rhizomatous by destroying this gene, its winter hardiness could be increased materially. Irradiation of sprigs could destroy this dominant gene and/or make other changes that could increase the winter hardiness of Coastcross-1.

On 21 June 1971, a mutation breeding project was initiated with
Coastcross-1 bermuda grass designed to increase winter hardiness by developing
winter-hardy mutants with rhizomes. Some 400 000 freshly cut green stems of
Coastcross-1 packed into 35 × 45 × 90 cm bales with a standard hay baler,
were transported 400 miles to the University of Tennessee Comparative Animal
Research Laboratory at Oak Ridge, Tennessee, and exposed to 7000 R. They
were then transported 150 miles to the Mountain Experiment Station at Blairsville,
Georgia, where they were broadcast and disced into the soil. The field in which
they were planted was sprayed immediately with 2,4-D to control weeds. Good
establishment was obtained and the presence of 25 chlorophyll-deficient mutants
proved that the irradiation treatment had been effective. A winter temperature
of −18°C (much lower than temperatures in most of Georgia) killed all control
plants but four tiny plants from irradiated material survived. These plants were
taken to Tifton where they were increased and planted in the field for further
evaluation. One survivor, labelled Coastcross-1-M3, developed a rhizome about
30 cm long, a characteristic not observed in the Coastcross-1 check but like Coastal.
This plant was increased by vegetative planting but failed to develop enough
rhizomes to be noticeably different from Coastcross-1 in this characteristic.

On 2 May 1974, Coastcross-1, Coastcross 1-M3 and Coastal bermuda grasses
were planted with 78 other bermuda grass hybrids in a 9 × 9 lattice square test
with five replications. Plots 3 × 5.3 m in size were given borders of bahia grass
to help keep them separated. All plots received 200 kg/ha of N and adequate
P and K each season. Strips of grass 0.6 m × 4.7 m were cut from each plot at
about six-week intervals in 1974, 1975 and 1976. The grass was weighed green
in the field and 350 g (approximate) samples were taken from each plot for dry
matter and in vitro dry matter digestibility (IVDMD) determinations.

During the 1974, 1975 and 1976 growing seasons following mild winters,
Coastcross-1 and Coastcross 1-M3 looked alike and did not differ in stand, early
vigour, dry matter yield or IVDMD (Table I). The Coastal bermuda grass check
(slower to establish) yielded less in 1974 and more in 1976 to give a three-year
average yield similar to that of Coastcross-1.

During the moderate winter of 1976–77, both Coastcross-1 and Coastcross
1-M3 suffered winter injury as evidenced by dry matter yields, only about half
those obtained from the Coastal bermuda check (Table II). However, Coastcross
1-M3 suffered less than Coastcross-1 and produced significantly more forage at the
6 June harvest. This difference in productivity carried over into the second
harvest but had disappeared by 1 September when the last cutting was made.
The data presented in Table II are considered evidence that Coastcross 1-M3 is a
mutant of Coastcross-1 with improved winter hardiness.

The severe winter of 1977–78 destroyed 98% of the grass in the clipped
plots of both Coastcross-1 and Coastcross 1-M3. Coastal bermuda grass suffered
very little loss of stand. Obviously, the small improvement in winter hardiness

TABLE I. THREE-YEAR DRY MATTER YIELD AND IN VITRO DRY MATTER DIGESTIBILITY (IVDMD) OF
THREE BERMUDA GRASSES DURING A PERIOD WITH MILD WINTERS AT TIFTON, GEORGIA

Variety	Dry matter (kg/ha)				Percentage IVDMD			
	1974	1975	1976	Ave.	1974	1975	1976	Ave.
Coastcross-1	8040	15 910	15 170	13 040	57.4	57.4	60.0	58.3
Coastcross 1-M3	8880	16 650	15 690	13 740	58.3	58.7	61.1	59.4
Coastal	6445	15 910	17 670	13 340	54.5	53.2	56.9	54.9
5% LSD	1350	1400	1740	2330	3.1	1.9	1.9	2.4

TABLE II. DRY MATTER YIELD OF THREE BERMUDA GRASSES
FOLLOWING A MODERATE WINTER AT TIFTON, GEORGIA

Variety	Dry matter (kg/ha)			
	6/6/77	20/7/77	1/9/77	Total
Coastcross-1	2090	2910	1150	6350
Coastcross 1-M3	3860	3690	1210	8760
Coastal	7720	5460	1870	15 050
5% LSD	1540	1040	480	2490

of Coastcross 1-M3 will not extend the zone of usefulness of Coastcross-1
very much. The small gain in winter hardiness exhibited by Coastcross 1-M3
suggests that there are several genes controlling the winter hardiness of well-
established Coastal bermuda grass.

For several years beginning in 1972 we have irradiated and planted more
than a million sprigs yearly of Coastcross-1 at Blairsville as in 1971. A series of
unusually mild winters delayed a satisfactory winter kill at Blairsville until the
winter of 1976–77. In the spring of 1977, we were able to find a few plants
that had survived the winter. Six of these have been increased and planted in a
replicated yield trial where they appear to be equal to Coastcross-1 in performance.
Unfortunately, they do not appear to be any more rhizomatous than Coastcross-1.

All the bermuda grasses used in our forage breeding programme are tetraploids
(2n = 36). Studies of rhizome development in other crosses involving the Kenya
bermudas indicate that more than one dominant gene is responsible for their lack
of rhizomes. This could explain our failure to see rhizomatous mutants in plants
from irradiated sprigs of Coastcross-1.

It is possible that the best 'mutant' from the 1971 irradiation that produced
some rhizomes and is a little more winter hardy may have had one less dominant
gene for absence of rhizomes or susceptibility to winter killing. If so, irradiation
of it might be more apt to produce rhizomatous or more winter-hardy mutants.
To test this hypothesis, we irradiated about a million sprigs from Coastcross 1-M3
in 1977. Plants from these sprigs produced some chlorophyll-deficient mutants,
but no mutants with extra rhizomes were found in the few surviving plants
observed in the early summer of 1978. To further test our hypothesis, we
irradiated and planted at Blairsville some 700 000 sprigs of Coastcross 1-M3 on
7 June 1979. These became well established and await a severe winter to eliminate
all but one or more plants that may be the winter-hardy mutants we seek.

Our results to date indicate that Coastcross-1 can retain its superior traits after exposure to gamma irradiation. A winter-hardy mutant with many rhizomes would have great economic value and would warrant all the effort put into this project.

REFERENCES

[1] BURTON, G.W., ROBINSON, B.P., The story behind Tifton 57 bermuda, South Turf Fdn. Bull. (Spring 1951) 3.
[2] ROBINSON, B.P., BURTON, G.W., Tiffine (Tifton 127) turf bermudagrass, U.S.G.A. Journal (June 1953) 30.
[3] BURTON, G.W., Tifgreen (Tifton 328) bermudagrass for golf greens, U.S.G.A. Green Sec. Rec. (May 1964) 11.
[4] BURTON, G.W., Tifway (Tifton 419) bermudagrass (Reg. No.7), Crop Sci. 6 (1966) 93.
[5] BURTON, G.W., Tifdwarf bermudagrass (Reg. No.8), Crop Sci. 6 (1966) 94.
[6] BURTON, G.W., "Improving turfgrasses", Turfgrass Science, Am. Soc. Agron., Madison, Wisconsin (1970) 410.
[7] POWELL, J.B., BURTON, G.W., YOUNG, J.R., Mutations induced in vegetatively propagated turf bermudagrasses by gamma radiation, Crop Sci. 14 (1974) 327.
[8] BURTON, G.W., Coastal bermudagrass, Ga. Coastal Plain Exp. Sta. Circular 10, Revised (1948).
[9] BURTON, G.W., Registration of Coastcross-1 bermudagrass (Reg. No.9), Crop Sci. 12 (1972) 125.

APOSPORIC EMBRYO SAC FORMATION IN INTERSPECIFIC *Festuca* HYBRIDS*

W. FRÖHLICH, K. GRÖBER, F. MATZK,
M. ZACHARIAS
Zentralinstitut für Genetik und
    Kulturpflanzenforschung,
Akademie der Wissenschaften der DDR,
Gatersleben,
German Democratic Republic

Abstract

APOSPORIC EMBRYO SAC FORMATION IN INTERSPECIFIC *Festuca* HYBRIDS.
   Embryological analyses were performed in F_1 hybrid plants of *Festuca pratensis* (2n=28) × *Festuca arundinacea* (2n=42), because in F_2 plants strikingly increased numbers of chromosomes were found. Just as in the sexual parental species the mature ovule of these F_1 hybrids is hemaeanatrope and crassi- to semi-crassinucellate. The formation of the embryo sac follows the *Polygonum* type, with the exception of an unreduced chromosome number in all nuclei. These results suggest that apomictic seed development can occur also after crossing between species which, as far as is known, are completely sexual.

INTRODUCTION

Experiments on the development of apomictic propagation have been reported earlier [1−3]. The results of hybridization between sexual species, including the genera *Lolium* and *Festuca*, with the aim of producing many different hybrids, were described. In addition, arguments were given to show that the voluminous material is an important prerequisite to the experiments. Up to the present nearly 2000 F_1 hybrids out of 50 different combinations have been produced. Some of these combinations were unknown in the literature. In about 20 crosses hybrids showed varying degrees of hybrid vigour regarding vegetative performance.

At present the type of propagation is being studied intensively in nine F_1 hybrids from five combinations. A complete embryological analysis has been carried out for three hybrid plants.

---

* Research carried out in co-operation with the IAEA under Research Contract No. 1995.

175

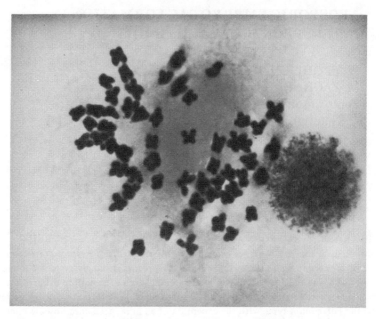

FIG.1.   First meiotic division in PMCs of the hybrid Festuca pratensis *(2n=28)* × Festuca arundinacea *(2n=42) in a plant with 2n=56 chromosomes from the* F_2 *generation.*

MATERIAL AND METHODS

Material from F_1 hybrids of the combination *Festuca pratensis* (2n=28) × *Festuca arundinacea* (2n=42) as well as from the parental species was fixed in different stages of development as whole spikelets or prepared in later stages under the stereomicroscope.

The material was fixed for 12–24 h in Carnoy's solution as follows: 6 parts alcohol (100%), 2.5 parts chloroform and 1 part acetic acid (99%). The fixative mixture was then washed out with alcohol (100%) and transported over alcohol (70%) in a 'Straßburger'-mixture. The usual procedures were used for embedding the material in paraffin wax (with the addition of ceresin and bees wax) or paraplast. Sections were cut between 6 and 15 μm, stained with haematoxylin and observed in euparal or caedax. Drawings were made from a projection of the original preparation during microscopic observation.

RESULTS

In offspring of some F_1 plants of the combination *Festuca pratensis* (2n=28) × *Festuca arundinacea* (2n=42) we observed F_2 plants with strikingly increased

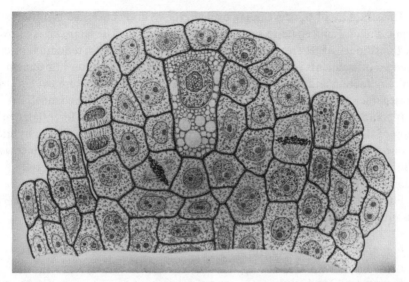

FIG.2. *Macrosporogenesis and development of the aposporous embryo sac. The formation of the legitimate EMC in the subepidermal cell layer.*

chromosome numbers. After uncontrolled pollination we found individuals with $2n = 56$ or more chromosomes (Fig. 1). It is possible that these F_2 plants could have been developed from an unreduced embryo sac (ES). Therefore an exact determination was made of the origin of the embryo sac mother cell (EMC) and the mature embryo sac (ES).

The generally uniform pattern of early inflorescence development in grasses enables a general description to be made of the stages in the transition from the vegetative to the floral apex. Sharman [4] has pointed out that the length of the vegetative apex of grasses is fairly constant for one species but variable between species. In general a short type of apex may be observed, as in *Avena, Secale* and *Triticum,* in an intermediate type such as *Festuca* and *Poa,* and in a long apex type represented by *Lolium* or *Melica.* In some cases it was observed that the length of the apex may be influenced by environmental conditions.

The earliest morphological change in the apex indicating the transition from the vegetative to the reproductive phase is the elongation of the apex and an increase in hypodermal cells in axils of the leaf primordia. In principle the top of the cell humps developing to a coenocarpe is derived almost entirely from the dermatogen. The integuments which in turn enclose the nucellus also arise through the division of the dermatogen. After the formation of integuments the ovular primordia grow crooked downwards, in the course of which they form a hemae-anatrope ovule.

The structure of the developing ovule in the hybrid concerned is of the bitegmic and crassinucellate to semi-crassinucellate type. The later stages of ovules show transitional stages from bi- to unitegmy, because there is a reduction of one of the two integuments. In the course of unitegmization the outer integument suffers a reduction and is compressed or resorbed fully in the mature ovule. Only one integument shapes the micropyle in the form of a micropylar channel.

It could be observed that very soon an archesporial cell is differentiated at the top of the ovule primordium. In the parental species as well as in the hybrid only one archesporial cell is formed, which is characterized by a striking increase in cells (Fig. 2). The archesporial cell transforms into the EMC without cell division.

For the formation of embryo sacs we tested the following species of the genus *Festuca:*

F. pratensis (2n=14 and 2n=28)
F. arundinacea (2n=42)
F. gigantea (2n=42)

In all cases meiosis in these sexual species is completely normal. In addition, hybrids of the following combinations were investigated:

F. pratensis (2n=28) $\times$ *F. arundinacea* (2n=42)
      (F_1 hybrid 2n=35)
F. pratensis (2n=28) $\times$ *F. gigantea* (2n=42)
      (F_1 hybrid 2n=35)

Offsprings of some F_1 hybrids, especially of the first combination, show the chromosome number of 2n=56. In these cases the mother plant possibly has been apomeiotic. Therefore a more exact analysis is being carried out to find out whether unreduced embryo sacs are produced.

In a series of distinct preparations we observed clearly that the legitimate EMC degenerates and one or more aposporous EMCs develop (Fig.3). The aposporous EMCs could be perceived as larger cells in the basic or lateral part of the nucellus. In some cases there are several cells arranged in this way, but as a rule only one cell was strikingly enlarged and operated as an aposporous initial cell (Fig. 4). Evidently there will be a competition between the aposporous EMCs in the ovule. This EMC, which is situated closer to the vessels and therefore also to the nutritive tract, will be the winner in this competition [5]. In the normal case this phenomenon obviously prevents polyembryony, in the combinations mentioned above.

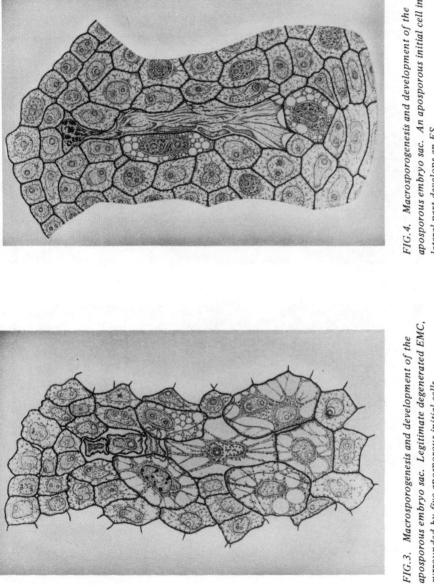

FIG. 4.   Macrosporogenesis and development of the aposporous embryo sac.   An aposporous initial cell in the lateral part develops an ES.

FIG. 3.   Macrosporogenesis and development of the aposporous embryo sac.   Legitimate degenerated EMC, surrounded by five aposporous initial cells.

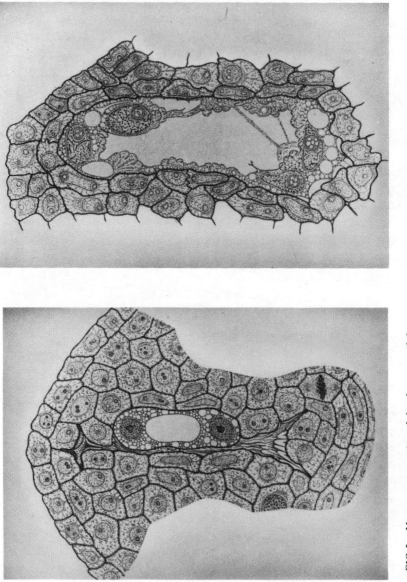

FIG. 6.   Macrosporogenesis and development of the aposporous embryo sac. Four-nucleate aposporous ES.

FIG. 5.   Macrosporogenesis and development of the aposporous embryo sac. Binucleate aposporous ES.

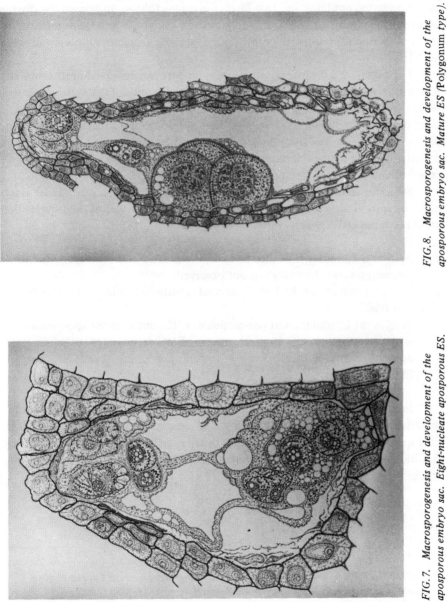

FIG. 8. Macrosporogenesis and development of the aposporous embryo sac. Mature ES (Polygonum type).

FIG. 7. Macrosporogenesis and development of the aposporous embryo sac. Eight-nucleate aposporous ES.

After a striking enlargement of the aposporous EMC the first mitotic division takes place followed by polarization. Then a central vacuole is formed between both nuclei which are situated at opposite poles (Fig. 5). Enlargement and vacuolation continue and two further divisions follow. In general the divisions occur synchronously in both cells and give rise to the four- and eight-nucleate stages (Figs 6 and 7). Now the central vacuole always separates both groups of nuclei.

The mature ES is piriform, its micropylar part is narrowed and its chalazal region enlarged. The egg apparatus consists of the egg cell and two synergids and one polar nucleus occurs on the chalazal pole. The fusion of the polar nuclei takes place near the egg apparatus, when the three antipodals are situated in the central part of the ES (Fig. 8). The completely formed ES is bipolar and, as in the sexual parents, belongs to the *Polygonum* type, with the exception of an unreduced chromosome number in all nuclei.

Stages of fertilization of the central nucleus were not observed. Later stages of ES or embryo development may indicate that in the ovules of the hybrids there are three types of ES development:

1. Neither a legitimate EMC nor an aposporous intial cell developed; the first degenerates and the latter was not observed.
2. The development of the EMC is normal, aposporous initial cells were not observed.
3. Besides the legitimate, but one-nucleate, EMC, one or more aposporous initial cells are present. In most cases the EMC degenerates at the one-nucleate stage and therefore one of the aposporous initial cells gives rise to ES development.

In future ES development will be studied in other clones from the hybrid material. The results of the embryological investigations show that aposporic ES formation takes place also in hybrids originating from crosses of sexual species. Additional hybrids from other combinations are available and we expect to find stocks with apomictic seed development.

REFERENCES

[1] GRÖBER, K., MATZK, F., ZACHARIAS, M., Untersuchungen zur Entwicklung der apomiktischen Fortpflanzungsweise der Futtergräsern. I. Art- und Gattungskreuzungen, Kulturpflanze 22 (1974) 159.
[2] GRÖBER, K., MATZK, F., ZACHARIAS, M., Untersuchungen zur Entwicklung der apomiktischen Fortpflanzungsweise der Futtergräsern. II. Hybrideffekt und Fertilität von Art- und Gattungsbastarden, Kulturpflanze 24 (1976) 349.

[3] GRÖBER, K., MATZK, F., ZACHARIAS, M., Untersuchungen zur Entwicklung der apomiktischen Fortpflanzungsweise der Futtergräsern. III. Beziehungen zwischen apomiktischer bzw. amphimiktischer Samenausbildung, vegetativer Leistung und Chromosomenzahl bei *Poa pratensis* L., Kulturpflanze **26** (1978) 303.

[4] SHARMAN, B.C., The biology and developmental morphology of the shoot apex in the Gramineae, New Phytol. **46** (1947) 20.

[5] NYGREN, A., "Apomixis in the angiosperms", Handbuch der Pflanzenphysiologie **18** (RUHLAND, W., Ed.), Springer Verlag, Berlin (1967) 551.

MUTATION INDUCTION AND ISOLATION IN POTATO THROUGH TRUE SEED AND TUBER MUTAGENESIS AND USE OF TISSUE CULTURE*

M.D. UPADHYA, M.J. ABRAHAM, B. DASS,
R. CHANDRA
Central Potato Research Institute,
Simla,
India

Abstract

MUTATION INDUCTION AND ISOLATION IN POTATO THROUGH TRUE SEED AND TUBER MUTAGENESIS AND USE OF TISSUE CULTURE.

Advance MV generation clones from hydrazine-sulphate-treated 'O.T' cultures have been field evaluated and 12 cultures have been selected for yield trials. One culture, DN-31-3, has been found to be day-neutral in its tuberization behaviour. Four JL/RA clones have been selected after a large-scale field trial. These clones are the selections from gamma-irradiated self seeds of Kufri Lauvkar (A-7416) and hybrid A-2235. Similarly 15 MV_3 clones have been selected from the populations raised from EMS- and DES-treated self seeds of A-2235. Day-neutral mutants have been selected from the fourth to seventh sprouts taken from EMS-treated tuber halves of Kufri Jyoti. From the sixth sprout harvest from EMS-treated Kufri Jyoti halves, one mutant, BCN-6-2, has been isolated which showed less than 30 cysts of *G. rostochiensis* in the MV_2 generation. This clone has been multiplied and made disease-free through apical meristem culture. Through the use of a new medium, PM-32, the plating efficiency of mechanically isolated single callus cells of dihaploid PH-258 is 30—35%. Nitsch's medium has been modified to formulate a new medium for direct embryogenesis in single callus cells of Phulwa. All stages up to the globular stage could be observed after five to six weeks of culture. Efforts were continued for the enzymatic isolation of single-leaf mesophyll cells from dihaploid PH-258. A new cell separation medium has been formulated which gives 80% viable cells. The LD^{50} and LD^{100} doses of EMS for the single callus cells of dihaploid PH-255 were found to be 500 ppm and 1000 ppm respectively.

INTRODUCTION

Potato, a vegetatively propagated crop plant, is ideally suited for improvement through mutation breeding, since any induced change can be maintained and multiplied by asexual propagation. However, to achieve the above aim it is important that (a) the highest mutation rate is obtained by using different plant parts/organs as units for mutagenesis, and (b) selection procedures should be

* Research supported by the IAEA under Research Contract No. 1369.

185

standardized for the isolation of desired mutations. For potato, it was suggested
by Nayar and Dayal [1] that taking successive crops of sprouts from mutagenized
tubers and planting them individually should be adopted to increase the induced
mutation frequency.

In this report results of experiments conducted for the induction and
isolation of useful mutations from successive harvests of sprouts from mutagenized
potato tubers as well as gamma-irradiated true self seeds are presented.

The successful regeneration of plants from potato callus as well as from
potato leaf mesophyll cell protoplasts has opened the way for the successful
utilization of tissue culture techniques for mutation induction and isolation at
the single cell level in potato. Therefore experiments were also carried out for
the standardization of cultural media/conditions for the isolation and culture of
single mesophyll cells from potato leaves and for increasing the planting efficiency
of single callus cells. Data are also presented on these studies.

MATERIAL AND METHODS

Assessment and evaluation of mutants isolated and selected during previous years

Day-neutral mutants obtained from the variety 'O.T' during 1972–73 [2],
and the yield behaviour reported earlier [3], were made disease-free and the tuber
material of subsequent MV generations was multiplied under the seed plot
technique at the Central Potato Research Station, Jullundur; assessments of
their yield and other characteristics were also made.

True self seeds of the hybrids A-7416 (Kufri Lauvkar), A-2235 and
EM/F-2128 were irradiated with 2, 4, 6, 8 and 10 krad of gamma irradiations from
a ^{60}Co source at the Indian Agricultural Research Institute, New Delhi[1] The
irradiated seeds were germinated in Petri dishes and transplanted in pots for
raising to maturity. The MV_1 tuber clones were grown in the field at the Central
Potato Research Station, Jullundur, and selected on the basis of yield and other
characters [3]. The selected clones have been multiplied using the seed plot
technique to maintain them in a disease-free condition. The selections have been
carried out in the successive MV generations up to MV_5. The seven selected clones
were then put on trial in a randomized block design replicated thrice, using
Kufri Chandramukhi, Kufri Sindhuri and Kufri Lauvkar as controls. This series
has been designated with JL/RA numbers. Similarly, 1000 true self seeds of
hybrid A-2235 were treated with 200 ppm of EMS and DES respectively for
4 and 8 h at room temperature after pre-soaking for 24 h at room temperature.
Treated and control seeds were then thoroughly washed in tap water and

---

[1] 1 rad = 1.00×10^{-2} Gy.

germinated in Petri dishes. Germinated seeds were transplanted in pots for growing to maturity. The MV_1 clones were grown in the field and selections were made on the basis of yield and other tuber characters. The selected clones have been given JL/RB numbers.

Use of tuber halves

Tuber halves of variety Kufri Jyoti were used for mutagenic treatment with the chemical mutagen EMS. The doses of EMS used were 100 and 500 ppm. The mutegenic treatment with the chemical mutagen was for 4 h at $25 \pm 2°C$. Following the mutagenic treatment, the tuber halves were washed in tap water and placed in trays in the dark at $25 \pm 2°C$ for sprouting. Successive sprouts were harvested, the sprouts were raised in pots to maturity and the selected MV_1 tubers and the later generation were grown in the field to maturity.

Screening of mutations for resistance to cyst nematodes

Sprouts from the 6th harvest of sprouts from treated tuber halves of Kufri Jyoti with 100 ppm and 500 ppm of EMS and controls were planted in plastic pots filled with soil infested with *Globodera rostochiensis* from Vijayanagaram Farm, Ootacamund. On the 30th day after planting, root ball counts for cysts were taken from each of the sproutlings from control and treated material.

Single cell isolation from callus and leaf mesophyll and culture

Different formulations of media for inorganic and organic constituents as well as various combinations of auxins (IAA and NAA) and cytokinins were tried with a view to:

(a) increasing the plating efficiency of single cells isolated from the callus of a dihaploid PH-258 of Up-to-Date;

(b) inducing direct embryogenesis from single cells from the callus of dihaploid PH-258;

(c) isolating living and viable single-leaf mesophyll cells from the leaves of the dihaploid PH-258, and culturing them to produce calli; different concentrations of various commercially available pectinases, singly or in combination, were tried in different formulations of cell separation media to isolate single mesophyll cells.

Since cells isolated from the callus of dihaploid PH-255 were treated with 10, 50, 100, 250, 500, 750 and 1000 ppm of EMS solution prepared in culture medium PM-32, for 3 h at $20 \pm 2°C$ in the dark. Viability of these single cells was checked using the FDR technique of Larkin [4], with a view to determining the LD^{50} and LD^{100} doses of EMS.

RESULTS AND DISCUSSION

Assessment and evaluation of selected mutants

Day-neutral mutants of variety 'O.T'

        The day-neutral mutants of variety 'O.T' which were selected during 1972–73
following the treatment of single eyes with 0.01M and 0.001M of hydrazine
sulphate were made disease-free and grown under the seed plot technique to keep
them disease-free. In total 61 clones were multiplied and maintained. During the
season 1978–79, these clones were assessed for their yield potential using the
duplicate-row test against standard checks such as Kufri Chandramukhi and Kufri
Sindhuri. These duplicate-row tests indicated that out of 61 mutant clones only
12 were superior in their yield and other tuber characters. These 12 selections
were grown during the 1979–80 season for large-scale multiplication so that in

TABLE I. THE SELECTED CLONES OF THE JL/RA SERIES IN THE
MV$_5$ AND MV$_6$ GENERATIONS OF GAMMA-IRRADIATED SELF SEEDS

Self seeds of genotype	Dose (krad)	MV$_5$ selections	MV$_6$ selections
A-7416	2	2	—
(Kufri Lauvkar)	4	6	JL/RA-148
	6	14	JL/RA-219, 237, 238, 261, 262, 284 and 357
	8	1	—
	10	—	—
A-2235	2	—	—
	4	—	—
	6	1	—
	8	3	JL/RA-643
	10	1	—
EM/F-2128	2	3	—
	4	—	—
	6	1	—
	8	1	—
	10	—	—

TABLE II. THE YIELD TRIAL DATA OF THE SEVEN JL/RA SELECTIONS
ALONG WITH THE CONTROLS

Sample No.	Clones	Yield (dt/ha)
1	JL/RA-148	325.7
2	JL/RA-219	297.9
3	JL/RA-237	280.7
4	JL/RA-262	362.9
5	JL/RA-284	369.6
6	JL/RA-357	263.9
7	JL/RA-643	376.8
8	Kufri Chandramukhi	263.9
9	Kufri Sindhuri	284.1
10	Kufri Lauvkar	278.8
	S.E. ±	19.5
	C.V. (0.05)	11.5

the 1980–81 season standard replicated trials could be laid out for their final
assessment.

These 61 clones were again evaluated for their day-neutral behaviour under
controlled conditions, both under short and long photoperiods. Only one clone,
DN-31-3, was found to be day-neutral, and it tuberized in 35 days under both
short (8 h) and long (continuous) photoperiods. The clone DN-31-3 was also
assessed for its yield in the duplicate row test at Kufri, and was found to produce
a yield comparable with Kufri Jyoti, a standard variety used as control.

JL/RA series

In each of the treatments with gamma irradiation ranging from 2, 4, 6, 8
and 10 krad, from each of the lots of true self seeds of the hybrids A-2235,
Kufri Lauvkar (A-7416) and EM/F-2128, the progenies were raised and the first
selection was made in the MV_1 generation. Data were presented in the report
of Kishore and co-workers [3]. The selected MV_1 clones were again grown in
the following season, and further selections made on the basis of yield and other
tuber characteristics. The MV_2 and successive generations have since been
grown and evaluated each year up to the MV_5 generation. The number of
selected clones in the MV_5 generation are presented in Table I, along with the
selections in the MV_6 generation. The nine MV_6 generation clones have been
put under large-scale multiplication. Out of these nine selected clones, seven

clones were put under trial using a randomized block design with three replications, each plot having six rows of 20 tubers each. The trial was harvested after 90 days of planting. The final yield trial data are given in Table II. It will be seen from Table II that the clones JL/RA-262, 284 and 643 out-yielded all the controls, i.e. Kufri Chandramukhi, Kufri Sindhuri and Kufri Lauvkar. However, the clone JL/RA-148 gave a yield at par with Kufri Sindhuri which was the highest yielding control variety. It is interesting to note that out of seven clones selected in the MV_6 generation, five are the self-irradiated progenies of A-7416 (Kufri Lauvkar) and only one of hybrid A-2235, with one from the hybrid EM/F-2128. Furthermore, out of the three clones out-yielding the controls, two, i.e. JL/RA-262 and 284, are the derivatives of Kufri Lauvkar, and the third one, JL/RA-643, is from hybrid A-2235.

True self seeds of hybrid A-2235 were treated with 200 ppm of EMS and DES respectively, and the seeds were germinated in Petri dishes. The seeds were then transplanted in pots for growing to maturity. The MV_1 clones were grown in the field and selections were made on the basis of yield and other plant characters. In the MV_2 generation a total of 48 clones were grown in duplicate rows and were assessed for their yield and other characteristics such as earliness. Out of these only 15 could be selected. The 15 MV_3 clones selected were grown in the following year under the seed plot technique for large-scale multiplication under disease-free conditions.

Use of tuber halves

Treatment of tuber halves of Kufri Jyoti with EMS

A total of seven successive harvests of sprouts was possible from the above material after which the mother tuber halves completely dried up. The details regarding the treatment and the screening until the third harvest of sprouts have been given previously [5].

Up to the screening of the third harvest of the sprouts, a few plants raised from sprouts of control tubers also showed induction of tubers when screened after 45 days of exposure under continuous light. Therefore the screening procedure was slightly modified and the plants raised from the 4th up to the 7th harvest of the sprouts were screened after 30 days of growth under continuous illumination. Under this modified procedure tuber induction was not observed in the control material, whereas in the treated material there was tuber induction. The details of the screening are given in Table III. It will be noticed that a maximum number of plants raised from the fourth harvest of sprouts from halves treated with 100 ppm of EMS showed tuber induction. The frequency of plants showing tuber induction showed a decline in the subsequent harvests of the sprouts.

TABLE III. RESULTS OF EXPERIMENTS WITH EMS-TREATED
KUFRI JYOTI

Harvest	Treatment	No. of sprouts obtained	No. of plants survived	No. of plants with tuber	No. of plants with stolon
Fourth	Control	409	401	–	33
	100 ppm	381	375	15	49
	500 ppm	373	364	8	36
Fifth	Control	304	259	–	18
	100 ppm	308	303	9	28
	500 ppm	311	299	5	21
Sixth	Control	99	94	–	9
	100 ppm	168	165	2	14
	500 ppm	178	169	2	16
Seventh	Control	31	30	–	6
	100 ppm	98	96	2	11
	500 ppm	116	116	5	16

TABLE IV. NUMBER OF PLANTS RAISED FROM THE SIXTH HARVEST
OF SPROUTS IN DIFFERENT CYST RANGES FROM KUFRI JYOTI
MUTAGENIZED WITH EMS, WHEN SCREENED AGAINST CYST NEMATODES

Treatment	Classes based on cyst numbers					No. of plants screened	Percentage survival
	0–30	31–60	61–90	91–120	121–150		
Control	–	23	43	32	–	100	98
100 ppm	12	18	23	29	8	100	90
500 ppm	9	17	28	26	7	100	87

Screening for mutations for resistance to cyst nematodes

Part of the material, i.e. 100 sprouts from the 6th harvest from the control
tuber halves of Kufri Jyoti, and tuber halves treated with 100 and 500 ppm of
EMS, were screened for resistance to the cyst nematode *G. rostochiensis*. The rest
of the sprouts from the 6th harvest were screened for tuber induction under a
long photoperiod. The sprouts were planted in plastic pots containing soil

TABLE V. DETAILS OF SCREENING AGAINST CYST NEMATODES IN THE MV_1 AND MV_2 GENERATIONS

Sample No.	Plant No.	Cyst No. MV_1	Cyst No. MV_2	Sample No.	Plant No.	Cyst No. MV_1	Cyst No. MV_2
1a	ACN-6-1	7	92	12	BCN-6-1	16	105
1b	ACN-6-1	7	88	13a	BCN-6-2	17	28
2a	ACN-6-2	12	105	13b	BCN-6-2	17	57
2b	ACN-6-2	12	84	14a	BCN-6-3	18	83
3	ACN-6-3	17	67	14b	BCN-6-3	18	93
4	ACN-6-4	18	77	15	BCN-6-4	20	108
5	ACN-6-5	19	81	16a	BCN-6-5	20	88
6a	ACN-6-6	19	94	16b	BCN-6-5	20	63
6b	ACN-6-6	19	57	17	BCN-6-6	22	52
6c	ACN-6-6	19	62	18a	BCN-6-7	22	71
7	ACN-6-7	20	128	18b	BCN-6-7	22	98
8	ACN-6-8	22	114	19	BCN-6-8	24	43
9a	ACN-6-9	23	48	20a	BCN-6-9	26	67
9b	ACN-6-9	23	71	20b	BCN-6-9	26	99
10a	ACN-6-10	27	81	21a	Control	–	84
10b	ACN-6-10	27	62	21b	Control	–	108
11a	ACN-6-11	29	36	21c	Control	–	92
11b	ACN-6-11	29	49	21d	Control	–	69

infested with cyst nematodes. On the 30th day after planting the root ball counts were made for cysts on each plant roots. Although every plant showed the presence of cysts on the root ball, there were differences in the number of cysts in the treated material compared with the control. Some plants showed a higher number of cysts in treated material compared with the control. The data are given in Table IV. A wide range in the number of cysts on the treated material clearly indicates a modification induced by the chemical mutagen EMS.

Tubers from the 20 plants which showed less than 30 cysts on the root ball of the treated material were retained for further confirmation of resistance. All the plants were given numbers starting with ACN-6 or BCN-6 (A indicating material treated with 100 ppm of EMS, and B indicating material treated with 500 ppm of EMS, CN indicating that the material was screened for resistance against the cyst nematode, and 6 indicating that the material was derived from the 6th harvest of sprouts). The details of the second screening are given in Table V. The second

TABLE VI. COMPOSITION OF MEDIUM PM-32

(a) Inorganic salts	mg/l
Na_2SO_4	200
KNO_3	500
KH_2PO_4	25
$CaCl_2 \cdot 2H_2O$	200
$MgSO_4 \cdot 7H_2O$	350
$ZnSO_4 \cdot 7H_2O$	5
$MnSO_4 \cdot H_2O$	20
H_3BO_3	5
KI	0.25
Na_2MoO_4	0.25
$CaCl_2 \cdot 6H_2O$	0.025
$CuSO_4 \cdot 5H_2O$	0.025
Na_2EDTA	3.5
$FeSO_4 \cdot 7H_2O$	2.5

(b) Organic constituents	
Nicotinic acid	5
Glycine	2
Thiamine-HCl	2
Pyridoxine-HCl	1
Inositol	100
Sucrose	20000
Glucose	10000

pH adjusted to 5.8 with INKOH before autoclaving at
15 lb/in for 15 min (1 in = 2.54×10^1 mm).

screening in the MV_2 generation has, however, indicated that only one plant,
BCN-6-2a, had less than 30 cysts, whereas the rest of the plants had a higher number
of cysts. This plant was retained and has been planted for multiplication in the
glasshouse. Simultaneously, apical meristems from this plant have been brought
from Ootacamund for the production of disease-free and cyst-nematode-free material
at Simla. Plants have been raised successfully and about 50 tubers are now
available for further multiplication during the current season at Kufri.

TABLE VII. COMPOSITION OF MEDIUM PCM-3

	mg/l
KNO_3	4000
$Mg(NO_3)_2 \cdot 6H_2O$	1000
$Ca(NO_3)_2 \cdot 4H_2O$	500
$(NH4)_2SO_4$	15000
KH_2PO_4	200
Na_2EDTA	35
$FeSO_4 \cdot 7H_2O$	25
Inositol	4000
Sucrose	20000
Glutamine	800
L-Serine	100

pH adjusted to 5.8 with INKOH and sterilized by cold
filtration.

Single cell isolation from callus and leaf mesophyll and culture

Increasing the plating efficiency of single cells isolated from dihaploid potato callus

When single cells from dihaploids are to be used for the induction and
isolation of mutants, it is very necessary that a suitable culture medium be
available where the plating efficiency of single cells reaches around 50%, if not
higher. Our experiments as well as those in other laboratories have shown the
plating efficiency of isolated cells from calli to be only around 10–15%. The
culture medium standardized earlier for the regeneration and maintenance of
potato callus, when used for plating single cells isolated from the callus of the
dihaploid PH-258 (dihaploid of Up-to-Date), attained a plating efficiency of
between 10–15% only. Therefore, efforts were directed to suitably modifying this
medium or to formulating a new medium with an improved plating efficiency.
With this in view, a new formulation of the culture medium PM-32 (Table VI) was
made, and single cells mechanically isolated from the callus of this dihaploid were
plated at a density of $1 \times 10^{3-4}$ cells/ml. Different concentrations of auxin and
cytokinin (6-BAP) were tried keeping the basic medium of PM-32. At the auxin
and cytokinin concentrations of 3 ppm of NAA and 0.25 ppm of 6-BAP, the plating
efficiency attainable is now 30–35%. Efforts are being made for further improvement in the plating efficiency.

TABLE VIII. COMPOSITION OF MEDIUM PM-2E$_2$

(a) Inorganic salts	mg/l
KNO$_3$	2000
(NH$_4$)$_2$SO$_4$	500
CaCl$_2$ · 2H$_2$O	500
MgSO$_4$ · 7H$_2$O	500
KH$_2$PO$_4$	100
MnSO$_4$ ·H$_2$O	25
ZnSO$_4$ · 7H$_2$O	10
H$_3$Bo$_3$	5
KI	0.25
Na$_2$MoO$_4$	0.25
CaCl$_2$ · 6H$_2$O	0.025
CuSO$_4$ · 5H$_2$O	0.025
Na$_2$EDTA	35
FeSO$_4$ · 7H$_2$O	25

(b) Organic constituents	
Sucrose	10 000
Glucose	10 000
m–inositol	100
L-glutamine	750
Glycine	2
Thiamine-HCl	2
GA3	0.1
6-BAP	0.25
NAA	0.25

pH adjusted to 5.8 with INKOH before autoclaving at 15 lb/in for 15 min. (1 in = 2.54 × 10^1 mm).

Direct embryogenesis from single cells

The successful culture of pollen grains of tobacco and tomato to produce embryos with the ultimate regeneration of haploid plants by Nitsch [6] has opened a way to initiate studies for standardization of cultural conditions for direct embryogenesis from single somatic cells. Such a technique will make it easy to recover mutagenic events at the single cell level directly as a plant. With

TABLE IX. COMPOSITION OF MEDIUM CSM-22

Sarbitol	0.35M
Succinic acid	250 mg/l
MES	400 mg/l
Na_2EDTA	37.5 mg/l
PVP-10	1000 mg/l
K_2SO_4	200 mg/l
KNO_3	50 mg/l
$MgSO_4 \cdot 6H_2O$	250 mg/l
Bovine serum albumina fraction (V)	1000 mg/l

pH of the medium adjusted to 5.8 with INKOH before
cold filter sterilization.

this in view, experiments were conducted to utilize Nitsch's medium to culture
single cells from the callus of Phulwa to initiate embryogenesis. However, these
initial efforts failed and the cells proliferated to produce loose cell masses
predominantly by unidirectional division to give long filamentous structures.
Therefore, Nitsch's medium was modified and different formulations were tried.
In one formulation, PCM-3 (Table VII), we observed globular-shaped structures
after 5–6 weeks of culture at 25°C. Careful examination of these cultures clearly
showed typical stages leading to embryogenesis as described by Street and
Withers [7] in somatic cell cultures of *Dacus carrota.*

These globular- or torpedo-shaped structures failed either to grow further
or differentiate into advanced stages of embryogenesis, even after prolonged
culture up to six months in fresh medium (PCM-3). Also, when these structures
were transferred to $PM-2E_2$ medium (Table VIII) or PCM-3 (Table VII) containing
different combinations of auxins and/or cytokinin concentrations, they failed
to grow further to differentiate into plantlets.

These studies are being continued to standardize a medium and cultural
conditions which would allow full embryo formation from single cells.

Single mesophyll cell isolation and culture

Since differentiation of callus into shoots/plantlets was achieved even in
the callus which had been grown up to eight transfers, efforts were directed
towards the isolation of single mesophyll cells in the live and active state, and
to culture them to produce calli. This approach has some advantages in handling
in comparison with the protoplasts.

With the possibility of raising a plant from a single cell, the method of exposing mutagen-treated plant cells to inhibitors and other toxic agents to select resistant mutants has a broad application.

There are a few reports of other plants which have been raised from single mesophyll cells. However, there is no report in the literature on the successful isolation of single mesophyll cells from potato leaf and culture to produce calli. We began work in this direction, and in the first stage towards large-scale isolation of viable and active potato leaf mesophyll cells we have been successful. The material used for these studies was the dihaploid of Up-to-Date, i.e. PH-258.

The procedure for single cell isolation is as follows. Surface sterilized 3rd and 4th leaves of actively growing plant were cut into 0.5 × 2.0 cm pieces and plasmolysed in 0.35M Sorbitol for 45 min. The plasmolyticum was removed and cold filter sterilized 0.5% solution of Macerase (Calliochem) in CSM-22 (Table IX) at pH .58 was added in the proportion of 60 ml solution for each gram of leaf tissue, in an Erlenmeyer flask of 250 ml capacity. The strips were incubated at 25°C for $1-1\frac{1}{2}$ h with occasional shaking. Single cells thus isolated were washed thrice with sterile Osmoticum (0.35M Sorbitol), and checked for viability using the FCR technique of Larkin [4] with fluorescine diacetate. The isolation medium was modified from the one reported earlier [5] because of failure in the culture of single mesophyll cells isolated using the earlier medium. The present isolation medium was formulated following the isolation medium of Servaites and Ogren [8] and Paul and Bassham [9] for the isolation of leaf mesophyll cells, and the leaf cell isolation procedure of Mawson and Colman [10]. Following the observations of Mawson and Colman, it was observed that the preplasmolysis of potato leaf strips for 45 min in 0.35M Sorbitol before enzyme treatment had improved the viability of the isolated cells, and there was minimum dislocation of the chloroplasts from their natural arrangement. Single mesophyll cell preparations having more than 80% viable cells were cultured in several media having different combinations of auxin, cytokinin and coconut milk including the media PM-32 (Table VI) which had proved to be excellent for culturing single cells from callus of dihaploid. Efforts have, however, failed so far to culture these cells successfully to initiate division leading to the development of callus.

Treatment of single callus cells with EMS

Single cells from the callus of dihaploid PH-255 were mechanically isolated in the PM-32 medium. The isolated single cells were checked for their viability using the FCR technique of Larkin and the viability was found to be more than 95%. For treating these cells EMS solutions were prepared in the PM-32 medium at concentrations of 10, 50, 100, 250, 500, 750 and 1000 ppm. The single cells were treated for three hours at 20 ± 2°C in the dark. The cells were then washed

three times with the medium PM-32 by centrifugation at 100 g for 2 min. Using
the FCR technique the viability was determined in each treatment taking three
samples from each and counting the proportion of cells showing fluorescence.
The LD^{50} and LD^{100} dosages were found to be 500 ppm and 1000 ppm
respectively. The cells were then cultured in PM-32 in 5 ml of the medium at
1×10^3 cells/ml. The calli from each of the cultures have been transferred to
solid medium for further growth and differentiation.

REFERENCES

[1] NAYAR, N.M., DAYAL, T.R., A method for securing high frequencies of induced
     mutations in the potato, Z. Pflanzenzuecht. **63** (1970) 155.
[2] UPADHYA, M.D., DAYAL, T.R., DEV, B., CHOUDHURI, V.P., SHARDA, R.T.,
     CHANDRA, R., "Chemical mutagenesis for day-neutral mutations in potato",
     Polyploidy and Induced Mutations in Plant Breeding (Proc. Meeting Bari, 1972),
     IAEA, Vienna (1974) 379.
[3] KISHORE, H., DAS, B., SUBRAMANYAM, K.N., CHANDRA, R., UPADHYA, M.D.,
     "Use of induced mutations for potato improvement", Improvement of Vegetatively
     Propagated Plants through Induced Mutations, Research Co-ordination Meeting Tokai, 197:
     Tech. Doc. 173, IAEA, Vienna (1975) 77 (unpublished).
[4] LARKIN, P.J., Purification and viability determinations of plant protoplasts, Planta
     128 (1976) 213.
[5] UPADHYA, M.D., CHANDRA, R., ABRAHAM, M.J., Mutation induction and isolation
     in potato through true seed and tuber mutagenesis and use of tissue culture technology ,
     Paper presented at the 3rd Research Co-ordination Meeting on Improvement of
     Vegetatively Propagated Plants and Tree Crops through Induced Mutations, Skierniewice,
     Poland, 22–26 May 1978 (unpublished).
[6] NITSCH, C., "Pollen culture – a new technique of mass production of haploid and
     homozygous plants", Haploids in Higher Plants, Advances and Potential   (KASHA, K.J.,
     Ed.), Univ. of Guelph Publication (1974) 123.
[7] STREET, H.E., WITHERS, L.A., "The anatomy of embryogenesis in culture", Tissue
     Culture and Plant Science (STREET, H.E., Ed.), Academic Press, London and New York
     (1974).
[8] SERVAITES, J.C., OGREN, W.L., Rapid isolation of mesophyll cells from leaves of
     soybean for photosynthetic studies, Plant Physiol. **59** (1977) 587.
[9] PAUL, J.S., BASSHAM, J.A., Maintenance of high photosynthetic rates in mesophyll
     cells isolated from *Papaver somniferum*, Plant Physiol. **60** (1977) 775.
[10] MAWSON, B., COLMAN, B., The effect of osmotic potential on photosynthetic rates
     of enzymatically isolated leaf mesophyll cells, Plant Physiol., Suppl. **59** (1977) 31.

STUDIES ON MUTATION BREEDING
IN SWEET POTATO (*Ipomoea batatas* (L.) Lam.) *

H. KUKIMURA
Institute of Radiation Breeding, NIAS, MAFF,
Ohmiya, Ibaraki-ken

Y. KOUYAMA
Faculty of Agriculture,
Mie University,
Kamihama, Tsu-shi, Mie-ken,
Japan

Abstract

STUDIES ON MUTATION BREEDING IN SWEET POTATO (*Ipomoea batatas* (L.) Lam.).
    Different genotypes were subjected to gamma rays and EMS to examine the effects on tuber skin colour mutation. Different mutation rates were obtained according to the genotypes. The gamma irradiation induced larger sector size of skin colour mutation than EMS. Gamma rays had an effect on inducing flowering in MV_1 which is utilized in cross breeding. Mutagenic treatment by gamma rays and EMS on the hybrid true seed which segregates in a Mendelian ratio for pigmentation in leaf, stem and tuber and for shape of leaf gave some bias to their segregation ratios. Effects of gamma-ray irradiation on quantitative characters, such as dry matter content and total sugar content in tubers, were also investigated in hybrid populations. The treatments enlarged genetic variations on both the characters, being more effective on total sugar content. Clonal progenies derived from mutagenic treatment by gamma rays and EI were investigated for their quantitative MV_4-MV_6 characters (tuber yield, dry matter content and total sugar content) in MV_4-MV_6. Heritabilities in a broad sense and phenotypic variances were estimated from the measurements on derivative strains obtained by random selection from mutagenic treatment plots. The genetic variations which transmitted clonally could be induced by mutagenic treatments. Artificial selection was effective only for tuber yield. It was considered that experiments on a larger scale were required for dry matter content and total sugar content. Mutant clones induced by gamma rays, EI and EMS were evaluated for their agronomic characters such as stem length, tuber yield, dry matter content and total sugar content. Mutant clones with short stem length decreased their tuber yield and vice versa. A few mutant clones were found to excel the originals in dry matter content and total sugar content. One sweet potato relative, *Ipomoea leucantha* Jacq., is very good material for studying the mechanisms of the sporophytic incompatibility system. It can be used for experiments on the artificial shift of the reproductive mode. From the experiments in applying gamma rays to incompatible combination of the plant, the conclusion was drawn that chronic irradiation during flower bud formation was effective in promoting pollen germination and recovering hybrid seed between incompatible clones, whereas acute irradiation was not. Mentor pollen killed by heavy doses of gamma rays was not effective for shifting the incompatibility. Some aspects of mutation breeding in sweet potato are also discussed.

* Research supported by the IAEA under Research Contract No. 1336.

199

1.   INTRODUCTION

Sweet potato, *Ipomoea batatas* (L.) Lam., was reported under cultivation
as early as 1605 in Ryukyu, Japan. The crop is adapted to the humid and
temperate regions of Japan, playing an important role as one of the main calorie
producing crops. The total planted annually to sweet potato has been approximately 55 000 ha in recent years, although more than 400 000 ha were planted
during the food crisis in the 1940s, the period of World War II.

Important features of the species are complete (though partially differentiated)
autohexaploidy and strong self- and cross-incompatibility. The plant propagates
vegetatively as well as by true seed, and is highly heterozygous. Genetical schemes
of its characters are not yet worked out. Neither geneticists nor breeders know
how yield of the crop is determined, how many genes are involved and how they
interact.

Breeding of sweet potato has been carried out by screening for promising
seedlings among hybrids produced by crossing between cross-compatible varieties.
The aims with regard to improvement are high productivity of tuber and starch,
disease and pest resistance, good appearance and edibility of the cooked tuber.
Spontaneous mutations have been utilized in the improvement of the crop.
Cultivar 'Beniaka' was found as a mutant from the original cultivar 'Yatsubusa'
in 1898 [1], and since then it has been one of the most popular cooking
type cultivars in Japan on account of its brilliant red skin colour and very good
edibility.

Since Miller [2] first reported on induced mutations by X-ray irradiation in
sweet potato, many reports have been published [3–11]. Gustafsson and Gadd [12]
reviewed the mutation in sweet potato up to 1964 and Kukimura [13] up to 1978.
Now in Japan sweet potato breeders and physiologists are examining induced
mutant strains with the possibility of release as commercial varieties.

We are working on the following three lines of mutation breeding in sweet
potato: (a) methodological mutagenesis, (b) direct use of mutants with improved
qualitative and quantitative characters, and (c) application of radiations on the
cross breeding. In the present report, recent research work carried out along the
above lines is described and mutation breeding in sweet potato is discussed.

2.   MUTAGENIC EFFECTS OF GAMMA RAYS AND EMS ON TUBER
      SKIN COLOUR

Somatic mutations in sweet potato are very frequently observed in the
natural condition and plenty of commercial varieties have been recovered in
farmers' fields as variants of superior characters such as shorter vines which save
labour, good cooking quality and more attractive shape and colouring of tuber

than the original cultivars. Among these characters, tuber skin colour is highly changeable and is easily detected in the mutation experiment. Mode of inherit-ance of the tuber skin colour is considered to be controlled by a few complementary alleles which interact epistatically [7]. Mutations of the character would be manifested by genetic background of the clones. The present experiment was undertaken to find out the effect of mutagens on the tuber skin colour in various genotypes of sweet potato.

2.1. Materials and methods

Cuttings of 37 clones, including cultivars and breeders' clones, were subjected to gamma-ray irradiation with a dose 15 kR at 375 R/h dose rate and EMS treatment of 2% aqueous suspension. In the EMS treatment, the materials were dipped into the suspension (approximately half of the shoots) for 1 h and were washed by tap water for 3 h afterwards. All the materials of MV_1 were planted in a test field by the usual method and manuring. Investigation on the change of tuber skin colour was carried out at the harvest of MV_1 plants. From among the harvested tubers some tubers were selected at random and were kept for the MV_2 generation check. The gamma-ray experiment was repeated for two years with identical experimental procedures. At the harvest of the MV_2 tubers from each clone, a similar investigation as with the MV_1 generation was made on the tuber skin colour.

2.2. Results

2.2.1. MV_1 generation

Seven clones from gamma-ray irradiation and seven clones from EMS treat-ment were found to mutate in their tuber skin colour in the MV_1. Only one clone, Chugoku 6, mutated by both the mutagens. The size of the mutated sector of the tuber skin colour tended to be larger from gamma rays than from EMS. In the following year, the gamma-ray experiment was repeated in the same way. The second gamma-ray experiment showed generally higher mutation frequencies in each genotype than the first experiment. In the second experiment, investigation on flowering was made and pollen fertility was scored. Induced flowers were normal and had fertile pollen. It is likely that the gamma irradiation forced flowers to initiate by depression of tuberization even if day length and temperature conditions were hardly suitable for flowering.

2.2.2. MV_2 generation

In the MV_2 generation some of the clones were observed to show tuber skin colour mutations which could not be detected in the MV_1 generation. Consequently,

TABLE I. EFFECTS OF GAMMA-RAY IRRADIATION (15 kR) AND EMS TREATMENT (2.0%) IN MV_2 ON MUTATION RATE OF TUBER SKIN COLOUR

Name of clone	Gamma rays				EMS			
	No. of plant	Mean size of mutated sector	Mutation rate based on plant unit	Mean size of mutated sector per total tuber	No. of plant	Mean size of mutated sector	Mutation rate based on plant unit	Mean size of mutated sector per total tuber
		cm	%	$\times 10^{-3}$cm		cm	%	$\times 10^{-3}$cm
Norin 1	144	0.393	9.7	23.1	213	0.347	9.1	7.2
Shiro Norin 1	194	-	-	-	107	0.014	0.9	14.3
Tachi Norin 1	86	0.462	18.6	30.8	236	0.601	15.3	16.3
Okinawa 100	78	0.090	3.8	30.1	740	0.661	16.4	28.7
Koganesengan	25	-	-	-	104	-	-	-
Tamayutaka	241	-	-	-	207	0.097	3.9	9.7
Gokokuimo	74	0.092	6.8	18.4	231	2.391	27.3	32.3
Okimasari	79	-	-	-	254	0.404	5.9	25.3
Fukuwase	92	0.079	7.6	13.1	267	0.343	7.9	12.7
Benisengan	60	-	-	-	11	-	-	-
Koukei 14	103	-	-	-	11	0.258	7.4	12.9
Norin 9	-	-	-	-	201	0.033	0.5	33.3
BR-13	71	0.023	2.8	11.5	246	0.575	10.2	18.0
BR-26	23	0.118	4.3	17.6	33	0.473	9.1	15.8
Chugoku 6	76	0.045	3.9	14.9	169	0.169	7.1	13.0
Kyushu 62	212	0.014	2.4	14.3	153	-	-	-
Kanto 33	-	-	-	-	40	-	-	-
Kanto 52	65	-	-	-	282	*	3.5	21.9
Kanto 60	52	-	-	-	205	0.047	1.5	15.6
Kanto 64	18	-	-	-	-	-	-	-
Kanto 66	44	0.021	2.3	21.1	20	-	-	-
Kanto 73	253	1.241	10.7	47.7	464	0.047	3.2	29.9
Kanto 74	31	-	-	-	429	5.938	24.5	45.0
Kanto 78	55	-	-	-	56	-	-	-
Kanto 79	65	-	-	-	68	0.166	11.8	20.8
Kanto 80	142	-	-	-	228	0.199	4.4	19.9
Kanto 82	119	-	-	-	13	-	-	-
Chikei 18-1072	73	-	-	-	104	0.012	1.0	11.8
Chikei 18-1706	547	-	-	-	343	1.000	0.3	-
Chikei 18-2981	14	-	-	-	39	-	-	-
Chikei 19-4881	20	-	-	-	31	0.262	9.7	87.4
Chikei 21-273	163	*	1.2	61.1	190	0.139	1.6	34.8
NR 510-59	163	-	-	-	-	-	-	-

* Not scored.

mutation was detected in 12 clones from gamma-ray irradiation and 22 clones from EMS treatment in the MV_2 generation. Mutation frequencies based on the plant unit were approximately similar in each genotype from both mutagenic treatments. The mean size of the mutated sector from gamma-ray irradiation was comparatively larger than from the EMS treatment, as in the MV_1 generation (Table I).

2.3. Discussion

Mutagenic agents can induce tuber skin colour mutations in sweet potato. Clowes [14] postulated the concept of promeristem based on studies of the root apex structure. Some plant species have the three layers of initial cell groups and the initials of the plerom and columella exist separately in apical regions, but periblem and the dermatogen have common initials [15]. If this is the case in sweet potato, the change of tuber skin colour by irradiation might be caused by rearrangement of the layers in the root initials.

Occurrence of skin colour mutations by gamma rays varied year to year to a considerable extent. It leads to the suggestion that the elimination of mutated tissues by the original tissue is affected by growing conditions of the plant.

In comparing the mutagens, the fact that EMS caused a smaller mean size in the mutated sector than gamma rays suggests that different mechanisms in penetration of mutagens to the portion of the cell which is involved were involved in the mutational event.

Materials employed here differ in their genotypes. The mutation rate based on plant unit ranged from 0 to 55.6%. Clones which did not show any changes in the tuber skin colour were considered to be homozygous concerning the tuber skin colour alleles as well as suppressors, if any. Furthermore, some clones, Norin 1 and Kanto 63, seemed to possess a differentially mutable potential. This might be explained by different genetic background and/or by the number of genes involved in the tuber skin colour expression.

In the MV_1 generation gamma-ray irradiation induced flowering by depression of tuberization. It can be utilized for artificial induction of flowers for cross breeding, instead of controlling the photoperiod or grafting on to non-tubering species of sweet potato relatives.

3.   EFFECTS OF MUTAGENIC AGENTS ON HYBRID SEED

Sweet potato is considered to be a complete [16] or partial [17] autohexaploid plant with high heterozygosity in its genetic nature. Genetical schemes of its characters are not fully worked out because of the difficulties in breeding tests since the species displays strong self- and cross-incompatibility. Although it is an autohexaploid plant, its hybrids will give segregation ratios not departing to any extent from 3:1 and 1:1 when crossed between simplex genotypes and between simplex and nulliplex genotype. Other than the disomic consequences, tetrasomic or hexasomic inheritance should be considered only when multiplex genotypes are concerned in the cross.

It is expected that mutagenic treatments can modify the genetic consequences involved in hybrid populations. It is of great importance to know whether such modification might be a help in practical breeding programmes. In the present experiment, morphological characteristics which are assumed to be controlled by the Mendelian gene system [7] and quantitative characteristics such as dry matter content which highly correlates with starch content and total sugar content in tubers are dealt with.

3.1. Materials and methods

F_1 hybrid seeds obtained from six cross combinations were irradiated by gamma rays with 40 kR dose at 1 kR/h dose rate. Seed was sown in plastic trays and survivals were transplanted to the test field. Tuber skin colour and stem colour were investigated during growth. Dry matter content and total sugar content of the tuber were determined after harvesting.

Throughout the experiments the gamma rays were applied to air-dried seed. Dry matter content and total sugar content of the tubers were determined by the method described elsewhere [9].

3.2. Results and discussion

3.2.1. Stem colour and tuber skin colour

Effects of gamma rays on growth of seedlings are shown in Table II. A dose as high as 40 kR gave very low survival rates with a range of 26–54%, contrasting with 54–84% in the control.

Some seedlings from irradiated seed developed cotyledons only and were lacking a terminal bud. In such seedlings, adventitious buds emerged from the hypocotyl and developed shoots 40–50 days after the normal terminal bud development.

In each cross combination, seedlings of the controls and irradiated plots segregated are shown in Table III. In some seedlings sectorial chimaera of tuber skin colour were occasionally detected. They were classified by predominant colouring. Segregations for seedlings with anthocyanin pigments in stem and tuber skin and without them are shown in Table III together with chi-square test. Inheritance of anthocyanin pigments expression in the higher plant is generally considered to follow the Mendelian gene system by control of a rather small number of genes. From the results in Table III, inheritance of tuber skin colour

TABLE II. GROWTH AFTER MUTAGENIC TREATMENT

No.	Cross Combination	Treatment	Number of Seed	Survival Rate	Rate of Albino Plant	Rate of Plant Unrecovered
				%	%	%
7138	Kanto 80 x	Control	150	76	1.8	1.3
	Koukei 14	40kR	500	26	13.8	46.2
7139	Kanto 80 x	Control	250	84	4.0	0.0
	Benisengan	40kR	500	41	1.0	46.1
7120	Fukukei D7-769	Control	100	54	7.4	1.9
	x Fukukei D7-54	40kR	500	31	3.9	30.5
7127	Kyukei 19-1001	Control	300	71	6.1	0.5
	x Koganesengan	30kR	800	54	8.4	11.2
7134	F61-20 x	Control	300	75	2.2	0.0
	NR 510-59	30kR	800	48	1.8	5.2
7131	NR 64-119 x	Control	300	62	0.0	0.0
	Tamayutaka	30kR	800	32	0.0	34.9

is explained by either monohybrid (3:1) or dihybrid (15:1) inheritance. The only exception was the cross NR 64-119 × Tamayutaka. Segregation of seedlings deviated significantly from the ratio 3:1. In all the cross combinations, the segregation ratio of tuber skin colour was not affected by gamma irradiation. On the contrary, gamma-ray irradiation affected the segregation for stem colour. Two cross combinations were investigated from stem colour segregation (Table IV). Frequencies of seedlings with anthocyanin pigment in the stem decreased when gamma irradiation was given. From the cross Kyukei 19-1001 × Koganesengan seedlings of stem with anthocyanin pigment and of stem without it segregated with a ratio of 1:3. This type of segregation can be explained by dihybrid inheritance and it is assumed that a dominant epistatic gene suppresses the basic anthocyanin pigment gene. The fact that mutagenic treatment induces tuber

TABLE III.　SEGREGATION RATIO AND CHI-SQUARE TEST

| No. | Cross Combination | Treatment | No. of Plants | | Total Number of Plants | Segregation Ratio | Chi-square |
			With Anthocyanin Pigment	Without Anthocyanin Pigment			
7138	Kanto 80 x	Control	179	17	196	15:1	1.965
	Koukei 14	40kR	196	14	210	"	0.062
7139	Kanto 80 x	Control	103	11	114	15:1	2.248
	Benisengan	40kR	126	5	131	"	1.324
7120	Fukukei D7-769	Control	26	16	42	3:1	3.841*
	x Fukukei D7-54	40kR	109	42	151	"	0.638
7127	Kyukei 19-1001	Control	133	44	177	3:1	0.013
	x Koganesengan	30kR	154	54	208	"	0.103
7134	F61-20 x	Control	50	70	120	1:1	3.333
	NR 510-59	30kR	59	79	138	"	2.899
7131	NR 64-119 x	Control	178	42	220 }	3:1	4.097*
	Tamayutaka	30kR	246	50	296 }	"	10.38**
						13:3	0.0168
						"	0.6461

*　Significant at 5% level　(Chi-square=3.84).
**　Significant at 1% level　(Chi-square=6.63).

skin colour change from colourless to red by anthocyanin pigment in sweet potato [7] is supported by this hypothesis. Mutagenic treatment might have reduced the frequency of dominant basic genes for skin pigmentation through mutations.

3.2.2.　Dry matter content and total sugar content

In each cross combination dry matter and total sugar content of the tuber were determined as shown in Table V. Total sugar content in the control from

TABLE IV. SEGREGATION FOR STEM COLOUR

No.	Cross Combination	Treatment	With Anthocyanin Pigment	Without Anthocyanin Pigment	Total Number of Plant	Segregation Ratio	Chi-square
7139	Kanto 80 x	Control	47	61	108	1:1	1.815
	Benisengan	40kR	45	84	129	"	9.211**
7127	Kyukei 19-1001	Control	36	146	182 ⎫	⎰1:3	1.900
	x Koganesengan	30kR	31	185	216 ⎭	⎱ "	13.06**
						⎰3:13	0.1268
						⎱ "	2.7926

*   Significant at 5% level (Chi-square=3.84).
** Significant at 1% level (Chi-square=6.63).

the cross NR 64-119 X Tamayutaka showed significant bias from the normal distribution curve. It suggests that such a cross combination contains highly effective genes on total sugar content. Among the six cross combinations employed here, seedlings from Kanto 80 X Koukei 14 showed the largest range of total sugar content, 0.478–5.175%.

Dry matter content was generally less affected in its variance and frequency distribution than total sugar content. Size of variance of dry matter content decreased significantly after gamma-ray irradiation in most of the cross combinations, resulting in greater decrease of C.V. than the controls. The C.V. of total sugar content is approximately two to three times larger than that of dry matter content. This indicates that the screening for seedlings of higher content of total sugar than the parents is possible.

Correlation between total sugar content and dry matter content is negative when materials of various genotypes from varieties were employed [18]. In the present experiment, only the control from F61-20 X NR 510–59 showed significant negative correlation, whereas it was not significant among its irradiated seedlings. More study is required with a sufficient number of seedlings in a plot.

TABLE V. STATISTICAL MEASUREMENTS OF DRY MATTER AND TOTAL SUGAR CONTENT

No. of cross combination	Treatment	Dry matter content							Total sugar content						
		No. of plant	Range	Mean	S.D.[a]	C.V.	Skewness	Kurtosis	No. of plant	Range	Mean	S.D.[a]	C.V.	Skewness	Kurtosis
7138	Control	192	19.3-37.8	26.14	4.58	12.5	0.65**	-0.50	193	0.478-3.623	1.869	0.497	26.6	0.15	0.10
	40kR	204	17.6-47.7	28.10	3.58**	12.7	0.17	0.02	204	0.905-5.175	1.946	0.554	28.5	0.64**	1.07**
7139	Control	111	18.0-40.0	26.76	4.47	16.7	0.47	-0.39	111	0.970-3.918	2.061	0.575	27.9	0.09	-0.04
	40kR	128	18.3-36.4	29.33	3.43**	11.7	-0.26	0.17	130	0.768-2.629	1.462	0.434	36.7	0.36	-0.46
7120	Control	40	19.8-35.4	26.65	4.11	15.4	0.28	-0.89	46	1.232-2.832	2.027	0.439	21.6	0.16	0.72
	40kR	155	19.8-36.0	28.54	3.13*	11.0	0.10	-0.24	154	0.824-4.487	2.110	0.559	26.5	0.14	1.43**
7127	Control	174	23.9-39.9	31.30	3.15	10.1	-0.12	-0.17	176	1.375-3.838	2.399	0.430	17.9	0.09	0.41
	30kR	207	20.2-39.3	31.41	3.40	11.0	-0.40*	0.29	210	0.777-4.585	2.270	0.556**	24.5	0.43**	0.71**
7134	Control	120	20.9-40.4	33.49	3.28	9.8	-1.02**	2.62**	131	1.062-4.790	2.360	0.561	23.8	0.22	2.30**
	30kR	131	20.6-42.0	32.44	3.60	11.1	-0.47*	0.71	125	1.610-4.985	2.716	0.575	21.2	0.45	0.67
7131	Control	220	23.5-39.8	29.76	2.81	9.4	0.13	0.06	218	1.073-4.777	2.194	0.463	21.1	0.72**	2.19**
	30kR	278	20.7-39.9	29.94	3.20*	10.7	-0.22	0.06	272	1.356-4.059	2.254	0.500	22.2	0.69**	0.43

* Significant at 5% level.   ** Significant at 1% level.
a) F-test by variance ratio to each control.

TABLE VI. PHENOTYPIC CORRELATION COEFFICIENT BETWEEN DRY MATTER CONTENT AND TOTAL SUGAR CONTENT

No.	Cross Combination	Treatment	r(DMC:TSC)
7138	Kanto 80 x Koukei 14	Control 40kR	0.0944 −0.0041
7139	Kanto 80 x Benisengan	Control 40kR	−0.0038 0.0057
7120	Fukukei D7-769 x Fukukei D7-54	Control 40kR	0.3557* −0.1243
7127	Kyukei 19-1001 x Koganesengan	Control 30kR	−0.0885 −0.1351
7134	F61-20 x NR 510-59	Control 30kR	−0.2282* 0.0442
7131	NR 64-119 Tamayutaka	Control 30kR	−0.0700 0.0001

* Significant at 5% level.

It suggests, however, that the relationship between characteristics may be affected by the combination of parental genotypes and by mutagenic treatment (Table VI).

Seedlings having stems with anthocyanin pigment decreased by gamma irradiation. It indicates that genes which control such a characteristic correlate with some factors concerning radiosensitivity. The evidence accumulated in potato [19] and in soybean [20] that the coloured plants with anthocyanin pigments are more radiosensitive than the colourless plant strongly supports this postulation.

4.   EFFECTS OF MUTAGENS ON QUANTITATIVE CHARACTERS

Only a few publications have dealt with the induced genetic variation in quantitative characters by mutagenic treatment in vegetatively propagated crops, whereas there are many publications on seed propagated crops [21–23, 25].

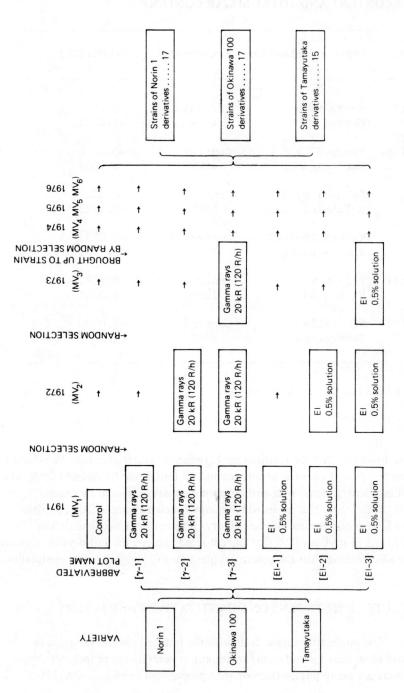

FIG.1. Experimental scheme: Effect of mutagens on quantitative characteristics.

It can be postulated that vegetatively propagated crops in general (fruit trees, ornamental trees and tuber crops) are large and are characterized by their low reproductivity. For these reasons it is difficult to extend the experimentation in such crops on a statistical basis with a sufficiently large number of sample plants. Also, in the crops there have been very few attempts to improve their quantitative characteristics including yielding capacity by mutation breeding. Under the circumstances we can find some reports employing sweet potato for its ease in handling and because its breeding targets are mostly in the improvement of quantitative characteristics such as high starch content and tuber yield.

Sakai [5] reported that he could screen sweet potato mutants of high dry matter content in which the starch content closely correlates by application of ^{32}P as a mutagenic agent. Marumine and Sakamoto [26] obtained a number of mutant clones from clonal progenies after ^{32}P and gamma-ray treatments. Among them, they recovered various mutant clones distinctive in high dry matter content, short vines and horizontal resistance against sweet potato sculf (*Monilochaeten infuscans* Ellis et Halsted).

Apart from sweet potato, in *Citrus* Sakai and co-workers [27] asserted that chronic gamma irradiation resulted in increased genetic variations in size of flower organs, such as petal, style, anther and stalk. To accumulate clonal variations in quantitative characteristics and to clarify the effects of the method of mutagenic treatment, recurrent treatment by gamma rays and ethyleneimine(EI) was carried out in clonal generations of sweet potato.

4.1. Materials and methods

Genetic measurements in sweet potato were estimated from variance analysis of several experiments in different growing conditions with 36–41 cultivars and breeders' strains. Calculation of heritability in a broad sense was made by the method described in the Appendix.

Three cultivars, Norin 1, Okinawa 100 and Tamayutaka, were employed in this experiment. Cuttings were subjected to mutagenic treatment with gamma rays and EI. The treatment scheme and experimental plots are illustrated in Fig.1. The gamma-ray dose was 20 kR at 120 kR/h dose rate and in the case of EI treatment, basal halves of cuttings were immersed in EI 0.5% aqueous solution for 3 h and were then washed down with tap water for 6 h. Forty plants per plot were raised in randomized blocks with five replications until the MV_3 generation. Plants in successive generations were brought up from seed tubers selected at random at the harvest. Plants which were considered to have mutated for their visible characteristics were all removed during the experiment from MV_1 to MV_4. In the MV_3 generation, five plants were taken at random from plants harvested in each plot and were brought up to strains originating from a single tuber in the MV_3 generation. After the MV_4 generation these strains were tested for their quantitative

TABLE VII. ESTIMATED HERITABILITY IN A GENERAL
SENSE OF AGRONOMIC CHARACTERISTICS IN
SWEET POTATO

Characteristics	Heritability(%)
Tuber weight per plant	21.8
Number of tubers per plant	32.0
Dry matter percentage of tubers	37.8–65.8
Total sugar content of tubers	9.6–14.3

characteristics of tuber, tuber yield, dry matter content and total sugar content.
During the MV_4 to the MV_6 generations the experimental design was the randomized
block method with five replications, each plot containing 40 plants. But, in earlier
generations than the MV_4, some plots decreased their plant number by shortage
of seed tubers. Thus 17, 17 and 15 derivative strains from Norin 1, Okinawa 100
and Tamayutaka respectively were established from each plot including the controls.
Consequently 333 derivative strains were tested in this experiment. Five plants
were taken from a plot as the sample for investigation of tuber yield, dry matter
content and total sugar content.

4.2. Results and discussion

Among the quantitative characters dry matter content showed a comparably
high heritability value in a general sense up to 65.8%, whereas the values of both
the tuber yield and total sugar content were outstandingly low as shown in Table V.
This suggests that the control of the latter two characteristics by artificial selection
is difficult.

In Norin 1, skin colour mutation was observed very frequently not only in
the MV_1 but also in successive generations without the treatment. Frequencies of
such mutations were scored in the MV_4 generation and were dependent on the
gamma-ray dosage (Table VIII). Other morphological mutations such as leaf shape,

TABLE VIII. MUTATION RATE IN TUBER SKIN COLOUR IN NORIN 1

Mutagenic treatment	Control	γ-1	γ-2	γ-3	EI-1
Mutation rate per tuber in the MV_4 (%)	0.6	1.1	5.7	5.2	1.9

internode length and flowering habit were observed to occur though at very low frequencies. These mutants were all removed so as not to contaminate seed tubers for the following generations since they derived from mutation of the major genes.

Sensitivity against the mutagenic treatment judged by vine length and tubering ability showed Norin 1 as the most sensitive and the remaining two cultivars at roughly the same level, Okinawa 100 being somewhat more sensitive than Tamayutaka.

Determination of dry matter content and total sugar content was carried out only in the MV_6 generation. Calculation of mean, phenotypic variance and heritability in a general sense was made in each mutagenic treatment plot in order to estimate whether the genetic variations were enlarged. Results are shown in Tables IX and X. As can be seen in Table IX, there were different responses to the mutagenic treatment among cultivars and among methods of treatment.

4.2.1. Tuber yield

Norin 1 derivatives showed statistically significant differences in variance between the control and the treatment plots only in the MV_6 generation, not in the MV_5. Mean values of the strains were lowered in the MV_6 generation. Values of heritability in a general sense were much higher than those of the controls in the MV_6 generation in all the cultivar derivatives with the mutagenic treatment. Okinawa 100 derivatives tended to increase mean values of the strains in the MV_6 generation but there were no significant variances by mutagenic treatment. Tamayutaka derivatives from the plots of double treatments with gamma rays (γ-2) and of single and triple treatments with EI (EI-1 and EI-3) showed significantly different variances from the controls. However, mean values were generally lower than the controls.

4.2.2. Dry matter content and total sugar content

The characteristics were investigated only in the MV_6 generation. In both characteristics, mean values decreased less than the controls though the differences

TABLE IX. GENETIC MEASUREMENTS OF TUBER CHARACTERISTICS IN CLONAL PROGENIES AFTER MUTAGENIC TREATMENT

[Tuber yield per plant]

Mutagenic treatment		Norin 1			Okinawa 100			Tamayutaka		
		Mean (g/plant)	δ^2	h^2 (%)	Mean (g/plant)	δ^2	h^2 (%)	Mean (g/plant)	δ^2	h^2 (%)
Control	MV_4	–	–	–	599	349	26	468	1294	17
	MV_5	202	388	30	384	385	41	336	1179	20
	MV_6	394	677	22	379	394	14	482	771	9
–1	MV_4	–	–	–	638	612	45	424	1710	34
	MV_5	241	489	37	348	616	43	306	1087	8
	MV_6	373	1387	62	392	732*	54	429	824	27
–2	MV_4	–	–	–	656	424	21	425	1371	18
	MV_5	279	458	33	369	613	62	312	2209**	54
	MV_6	357	913	42	444	374	10	437	1642**	63
–3	MV_4	–	–	–	666	452	25	354	2686**	58
	MV_5	285	450	33	373	296	22	362	1715**	41
	MV_6	374	1035*	49	427	469	28	441	824	27
EI-1	MV_4	–	–	–	583	696	52	418	1498	25
	MV_5	330	481	34	433	339	32	340	1674**	40
	MV_6	462	2199**	76	415	599	44	453	1075**	44

		(%)		(%)	(%)		(%)	(%)		(%)
EI-2	MV$_4$	—	—	—	649	369	9	390	1807*	38
	MV$_5$	—	—	—	434	370	38	333	1598*	37
	MV$_6$	—	—	—	432	480	30	446	749	20
EI-3	MV$_4$	—	—	—	604	397	15	379	2171*	48
	MV$_5$	—	—	—	396	335	31	336	1774**	43
	MV$_6$	—	—	—	463	607	44	429	894*	33

[Dry matter content and total sugar content in MV$_6$]

		(%)		(%)	(%)		(%)	(%)		(%)
Control	DMC	35.8	4.11	11	29.2	5.28	4	31.8	3.11	22
	TSC	3.37	0.283	23	2.85	0.178	13	2.84	0.286	8
-1	DMC	36.2	4.87	15	28.7	8.98*	39	30.7	6.32**	40
	TSC	3.45	0.900**	76	2.82	0.441**	26	2.71	0.442**	40
-2	DMC	35.1	9.92**	45	29.1	13.31**	59	31.8	7.22**	47
	TSC	3.25	0.571**	62	2.76	0.303*	19	2.63	0.363**	27
-3	DMC	35.3	28.35**	81	30.1	12.46**	56	31.8	6.94**	45
	TSC	3.58	0.473*	41	2.81	0.632**	48	2.80	0.645**	43
EI-1	DMC	35.4	16.79**	68	29.2	9.19*	40	31.8	5.82*	35
	TSC	3.06	0.543*	60	2.93	0.744**	56	2.89	0.647**	59
EI-2	DMC	—	—	—	28.1	5.65	3	31.1	4.73	20
	TSC	—	—	—	2.83	0.442**	22	2.81	0.482**	45
EI-3	DMC	—	—	—	29.2	4.33	27	31.8	6.94**	45
	TSC	—	—	—	2.91	0.827**	60	2.75	0.291*	

* Significantly different from the control at 5% level.
** Significantly different from the control at 1% level.

TABLE X. MEAN VALUE OF MUTAGENIC TREATMENT BASED ON RESULTS FROM TABLE III

	Tuber yield								Dry matter % (MV6)			Total sugar % (MV6)		
	(MV4)		(MV5)			(MV6)								
	Okinawa 100	Tamayutaka	Norin I	Okinawa 100	Tamayutaka	Norin I	Okinawa 100	Tamayutaka	Norin I	Okinawa 100	Tamayutaka	Norin I	Okinawa 100	Tamayutaka
Mean of derivative strains*														
Control	100	100	100	100	100	100	100	100	100	100	100	100	100	100
Mean of γ-ray treated plots	109	86	132	95	97	94	111	90	99	100	99	102	98	96
Mean of EI plots	102	84	163**	110	100	117**	115	92	99**	99	99	91**	101	99
Total mean of γ-ray treated & EI	106	85	140	102	99	100	113	91	99	100	99	99	100	98
Heritability in broad sense														
Control	26	17	30	41	20	22	14	9	11	4	22	23	13	8
Mean of γ-ray treated plots	30	37	34	42	34	51	31	39	47	51	44	60	31	37
Mean of EI plots	25	37	34**	34	40	76**	39	32	60**	23	33	60**	46	38
Total mean of γ-ray treated & EI	28	37	34	48	37	63	35	36	50	37	39	60	39	38

*　Means are shown by indices, taking the controls as 100.

were small. Difference in variances between the control and the derivatives by the mutagenic treatment were statistically significant at the 1% level in nearly all cases. In other words, despite a slight downward shift in the mean values, the variances in these characteristics were greatly increased by the mutagenic treatment. Consequently, the heritability in a general sense in the plots of mutagenic treatment increased more than the controls in all the cultivars. These facts indicate that the genetic variations in these quantitative characteristics transmitted by clonal propagation increased more in the derivatives from the mutagenic treatment than in the controls.

Variances of dry matter content in Norin 1 derivatives by gamma irradiation in the MV_6 generation and tuber yield of Tamayutaka derivatives by EI in the MV_4 increased as the treatment was repeated. However, other variances in total sugar content in Norin 1 derivatives in the MV_6 by gamma irradiation and in dry matter content of Okinawa 100 derivatives in the MV_6 by EI treatment decreased as the treatment was repeated. There was no fixed tendency in the relationship between increment of the variances and the method of treatment. Furthermore, there were no clear differences between the two mutagens in the effect on the genetic measurements of the three characteristics, though the means of derivatives in the three characteristics after gamma irradiation were slightly lower than those after EI treatment.

It is concluded that in sweet potato the genetic variations transmitted by clonal propagation can be induced by mutagenic treatment. These variations in tuber yield, dry matter content and total sugar content which are controlled by the polygenic system can be controlled by artificial selection.

5.   SELECTION EXPERIMENT ON QUANTITATIVE CHARACTERISTICS
     AFTER MUTAGENIC TREATMENT

A selection experiment was conducted following the results described above (see Section 4) in order to find out the effects on quantitative characteristics.

5.1. Materials and methods

Cuttings of cultivars Norin 1, Okinawa 100 and Tamayutaka were subjected to mutagenic treatment with gamma rays and EI. The treatment scheme of the experimental plot is shown in Fig. 2. The gamma-ray dose was 20 kR/h and EI was applied with 0.5% aqueous solution. The number of plants in the MV_1 was 120 in each treatment and thereafter until the generation in which selection was carried out 40 plants per plot were raised in randomized blocks with five replications. Plants in the generation following each treatment were

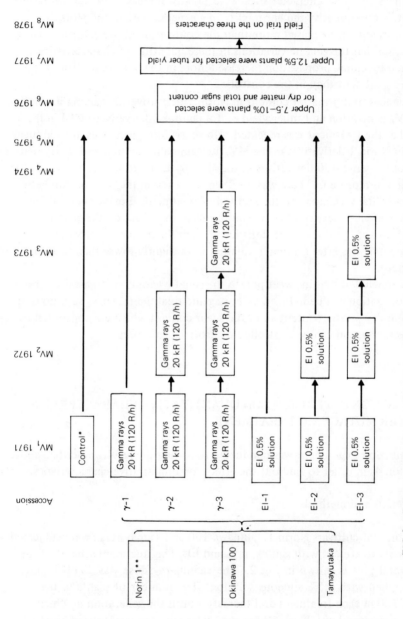

FIG. 2. Experimental scheme: Selection experiment on quantitative characteristics.
* Selected clones from the controls were applied for estimating the confidence level at the 5% level.
** In Norin 1, treatment with EI-1 and EI-2 was not carried out.

brought up from seed tubers taken at random, eliminating visible mutants. The upper 7.5−10% of plants was selected for dry matter content and for total sugar content from the results of its determination at the MV_6 harvest and the upper 12.5% of plants was also selected for tuber yield at the MV_7 harvest. The selections were carried out in every replicated plot.

In the MV_8 generation a field trial was carried out on the three characteristics: dry matter content, total sugar content and tuber yield. The experimental design employed in the MV_8 was the standard block method with three replications of five plants per plot. Tuber yield was weighed in the field immediately after lifting from the soil.

5.2. Results and discussion

Statistical measurements of the three quantitative characters from the field trial of selected clones in the MV_8 are shown in Table XI. Confidence intervals which were calculated from the selected clones out of the controls at the 5% level are shown. Their variations should have originated from the environmental effects only. For each mutagenic treatment the number of clones which exceeded the confidence interval of each original control is shown. In tuber yield the effect of the selection was significant in Norin 1 derivatives. As is clear from Table XI, nearly half the selected clones exceeded the confidence interval of the control, the mean of tuber yield per plant in the selected clones of mutagenic treatment being in the range 600−700 g which differed significantly from 567 g of the control. It was approximately 115% of the control.

On the other hand, in the other cultivar derivatives, Okinawa 100 and Tamayutaka, there were no significant increases by the selection after mutagenic treatments. Very few clones exceeded the controls, and the average tuber yield was less than the control, though not significantly so. This was presumably due to the smaller number of plants which were subjected to the selection in Okinawa 100 and to insufficient dosage to induce any genetic change because of low radiosensitivity in Tamayutaka [28]. Since the cultivar Norin 1 is one of the most volatile cultivars in visible characteristics, it may be highly mutable also in quantitative characteristics.

To determine dry matter content and total sugar content requires a lengthy processing of tuber samples and chemical analysis. The clones selected from the investigation of both characteristics on the MV_6 harvest were tested in the MV_8 generation. The effect of selection on both characteristics was not significant except in one case of a single treatment with EI on Tamayutaka. In almost all the mutagenic treatments of every cultivar, the mean of selected clones did not excel the confidence intervals of the controls. It suggests that there may be some question in selection intensity. A study on a larger scale is required to evaluate any effects of selection in such characteristics.

TABLE XI. STATISTICAL MEASUREMENTS OF QUANTITATIVE CHARACTERISTICS IN THE MV$_8$ GENERATION AFTER SELECTION

Mutagenic treatment	Tuber yield (kg/plant)						Dry matter content (%)						Total sugar content (%)					
	No. of clones	Range	Mean	Variance	Confidence interval a)	(No. of clones)	No. of clones	Range	Mean	Variance	Confidence interval	(No. of clones)	No. of clones	Range	Mean	Variance	Confidence interval	(No. of clones)
[Norin 1]																		
Control	40	.255-.920	.567	.0186	.523-.610		20	33.8-37.5	35.04	1.06	34.56-35.52		20	4.19-5.15	4.51	.074	4.38-4.63	
γ-1	38	.255-1.070	.701**	.0322		(22)	20	32.1-38.5	33.78	2.53		(2)	20	4.10-5.12	4.44	.092		(3)
γ-2	40	.263-1.038	.637*	.0321		(21)	20	33.7-37.0	35.09	0.91		(6)	20	4.00-5.01	4.45	.099		(6)
γ-3	32	.176-1.040	.613	.0346		(15)	20	32.2-37.7	34.01*	2.55		(3)	20	4.34-5.75	4.66	.122		(8)
EI-1	38	.210-1.038	.637*	.0192		(25)	20	34.2-38.0	35.38	1.18		(6)	20	4.18-5.93	4.58	.187*		(6)
[Okinawa 100]																		
Control	26	.373-1.356	.833	.0721	.742-.941		15	26.3-28.5	27.53	0.65	27.08-27.97		15	3.90-5.57	4.39	.210	4.13-4.64	
γ-1	24	.503-1.230	.741	.0312		(3)	15	26.5-31.8	28.32	2.34*		(8)	15	3.49-6.62	4.04	.569		(1)
γ-2	24	.433-1.327	.785	.0410		(4)	15	26.0-29.5	27.50	1.01		(3)	15	3.59-4.60	3.97*	.107		(0)
γ-3	21	.397-1.523	.848	.0956		(9)	15	26.3-33.0	28.28	3.12*		(9)	15	3.60-4.53	3.99*	.072		(0)
EI-1	29	.310-1.210	.559*	.0356		(1)	15	25.5-32.0	27.31	1.82		(4)	15	3.77-6.04	4.31	.323		(2)
EI-2	25	.407-1.280	.746	.0483		(4)	15	25.3-29.0	26.77	1.46		(3)	15	3.83-8.34	4.43	1.200**		(1)
EI-3	23	.463-1.317	.747	.0415		(2)	15	26.0-30.7	28.27	1.73		(8)	15	3.45-4.36	3.91**	.110		(0)
[Tamayutaka]																		
Control	44	.330-1.080	.718	.0330	.662-.773		20	31.5-35.7	32.73	1.31	32.19-33.26		20	3.38-4.55	3.81	.123	3.65-3.97	
γ-1	40	.503-1.280	.725	.0211		(12)	20	30.9-35.4	32.41	1.72		(6)	20	3.41-4.86	3.81	.149		(5)
γ-2	40	.347-1.290	.641	.0383		(8)	20	30.8-34.4	32.13	1.36		(3)	20	3.23-4.80	3.67	.071		(3)
γ-3	39	.260-1.076	.642	.0275		(6)	20	29.5-34.0	31.75*	2.10		(4)	20	3.12-3.87	3.46**	.220		(0)
EI-1	50	.347-1.167	.684	.0556		(15)	20	31.1-36.3	32.86	2.46		(5)	20	3.75-5.91	4.40**	.220		(18)
EI-2	47	.360-.990	.670	.0186		(12)	20	31.7-34.7	32.97	0.93		(8)	20	3.50-5.38	3.98	.199		(8)
EI-3	49	.327-.947	.593	.0229		(6)	20	31.7-37.1	32.54	2.40		(11)	20	3.32-4.57	3.76	.137		(4)

* Significantly different from each control at 5% level.
** Significantly different from each control at 1% level.
a) Figures in parentheses are the number of clones above the confidence interval (at 5% level) of control.

Though the influence of mutagen difference on the effect of the selection was not definitely clear, there was a tendency for gamma rays to produce higher mean values in tuber yield in selected clones than the EI treatment. Differences in mean tuber weight per plant between gamma-ray and EI treatment were 13, 107 and 20 g in Norin 1, Okinawa 100 and Tamayutaka respectively. In Okinawa 100, gamma rays were more effective in inducing clones which exceeded the controls than EI treatment through the three characteristics, whereas in Tamayutaka EI was more effective. This suggests the low radiosensitivity of Tamayutaka, as mentioned above. The cultivar may need a much higher dosage to reach the level of the other cultivars.

The effect of recurrent treatment on the mutagens was not clear, but there was a reduction of the number of clones which exceeded the tuber yield of the controls. The mean value of selected clones rather decreased when the mutagenic treatment was repeated, as shown in the treatment plots of gamma-ray irradiation on Norin 1 and Tamayutaka and of EI treatment on Tamayutaka. It is considered that recurrent treatment of mutagens may cause some damage which is transmitted somatically to the following generation. The transmission of radiation damage takes place even later than in MV_2 in potato [9], and a similar situation is also observed in sweet potato. In dry matter content and total sugar content, such a decrease was not detected and the clones with higher yield were found more frequently in the recurrent treatment by EI.

As described above, only the tuber yield responded positively to the selection. The effects of mutagen differences and of recurrent treatment must be investigated more thoroughly. Since the test of selected clones in the field only lasted one year in the present experiment, it needs more repetition to draw precise conclusions.

6. AGRONOMIC CHARACTERISTICS OF INDUCED MUTANT CLONES

According to the Agriculture Census of Japan in 1977 [29], working hours per one-tenth of a hectare in sweet potato, potato, wheat and soybean cultivation are 80.2, 15.8, 22.6 and 13.9 h respectively. In sweet potato cultivation, additional farming processes are required such as growing seedlings in the nursery, transplanting by hand and removing vines just before the harvest which are not necessary in the other field crops. Breeders therefore are concentrating their efforts on improving this situation. By using true seed sowing or direct planting of seed tubers they save working-hours. Short-vined clones and dwarf-type clones are required to cut down labour at the harvest.

A number of mutant clones of sweet potato were obtained through methodological studies of mutation breeding at our Institute. Among them are some clones which might attract the breeder's attention from the point of view of practical cultivation.

TABLE XII. AGRONOMIC CHARACTERISTICS IN INDUCED MUTANT CLONES

Clones	Stem length	No. of nodes	Weight of tuber (g/plant)	No. of tubers	Dry matter content(%)	Total sugar content(%)	Remarks *	Dosage
[Mutant clones of vine length]								
Norin 1	229(100)	74(100)	475	2.8	29.75	5.381	Original	
N1-GM-9	143 (62)	37 (50)	400	2.6	35.76	6.138	WSM	15kR
N1-GM-13	195 (85)	60 (81)	450	2.4	33.46	5.296	"	28"
N1-GM-14	196 (91)	54 (73)	524	2.6	31.74	5.184	"	20"
N1-GM-23	208 (83)	70 (95)	420	2.6	30.26	5.456	"	4"
N1-GM-27	191 (83)	51 (69)	242	1.0	31.92	5.307	"	28"
Tamayutaka	140(100)	45(100)	548	2.4	30.30	4.135	Original	
TAM-GM-4	80 (57)	25 (56)	461	2.1	28.48	4.594	WSM	15kR
TAM-GM-6	56 (40)	20 (44)	478	2.6	30.40	4.653	RSM	"
TAM-GM-13	73 (52)	23 (51)	398	1.8	28.76	3.579	WSM	"
Tachi Norin 1	31(100)	32(100)	340	4.3	29.77	5.339	Original	
TN1-GM-3	173(558)	46(144)	630	3.0	28.82	4.231	BRSM	4kR
TN1-GM-4	118(381)	38(119)	680	2.8	28.50	4.620	SRSM EMS 2%	
TN1-GM-10	65(210)	25 (78)	320	2.3	30.45	6.694	WSM	"
TN1-GM-11	110(355)	38(119)	604	3.7	27.00	4.737	"	15kR
TN1-GM-12	148(477)	46(144)	870	3.4	30.25	5.134	"	"
TN1-GM-13	85(274)	27 (84)	420	3.7	26.13	4.516	"	"
Okinawa 100	245(100)	42(100)	660	5.7	31.62	4.542	Original	
O-100-GM-2	93 (38)	24 (57)	420	3.3	26.66	5.961	WSM EMS 2%	
O-100-GM-3	48 (20)	24 (57)	540	5.0	32.04	5.424	RSM 15kR	
O-100-GM-4	55 (22)	21 (50)	465	4.3	27.63	4.872	WSM EI 0.5%	
O-100-GM-6	92 (38)	36 (86)	520	4.3	29.38	6.224	" EI 0.3%	
O-100-GM-7	43 (18)	20 (48)	450	4.0	31.34	5.754	PGM EMS 2%	
O-100-GM-9	46 (19)	42(100)	224	3.0	25.36	5.863	WSM EI 0.5%	
Kanto 63	141(100)	39(100)	650	8.0	30.24	5.881	Original	
K63-GM-1	78 (55)	23 (59)	436	5.0	29.95	4.218	RSM 15kR	
K63-GM-3	76 (54)	28 (72)	490	5.6	30.13	5.450	" "	
K63-GM-6	83 (59)	30 (77)	480	6.5	29.98	5.814	" "	
K63-GM-7	83 (59)	29 (74)	468	1.9	31.05	5.305	" "	
K63-GM-9	91 (65)	40(103)	518	6.0	35.19	4.465	" "	
[Mutant clones of DMC and TSC]								
Norin 1	229	74	475	2.8	29.75(100)	5.381(100)	Original	
N1-GM-3	223	61	494	1.4	33.08(111)	8.439(157)	GSM 50kR	
N1-GM-6	211	67	350	2.8	32.54(109)	6.549(122)	" EI 0.5%	
N1-GM-9	143	37	400	2.6	35.76(120)	6.138(114)	" 15kR	
N1-GM-11	281	69	588	2.6	34.80(117)	5.165 (96)	PRSM 28"	
N1-GM-20	205	62	380	3.0	29.85(100)	6.926(129)	WRSM 20"	
N1-GM-30	236	63	364	2.4	27.17 (91)	6.937(129)	WSM 15"	
N1-GM-43	263	60	482	2.2	34.22(115)	5.380(100)	" 8"	
N1-GM-46	281	62	456	4.0	34.42(116)	5.638(105)	GSM 38"	
Tamayutaka	140	45	548	2.4	30.30(100)	4.135(100)	Original	
TAM-GM-2	96	28	512	2.2	31.70(105)	5.263(127)	WSM 15kR	
TAM-GM-3	105	32	446	2.4	32.74(108)	4.354(105)	RSM EI 0.5%	
TAM-GM-8	113	33	570	2.9	28.50 (94)	4.950(120)	WSM 15kR	
TAM-GM-18	103	37	420	1.8	31.40(104)	5.984(148)	" EI 0.5%	

* WSM=White tuber skin mutant, RSM=Red tuber skin mutant, BRSM=Brown tuber skin mutant, SRSM=Scarlet tuber skin mutant, PGM=Pale green leaf mutant, PRSM=Pale red tuber skin mutant, WRSM=Red tuber skin mutant reversed from WSM, GSM=Green stem mutant. Original Norin 1=Red tuber skin, Tamayutaka=Red at top and tail of tuber otherwise white, Tachi Norin 1=Red tuber skin, Okinawa 100=Red tuber skin, Kanto 63=White tuber skin.

6.1. Materials and methods

Cultivars Norin 1, Tachi-Norin 1, Tamayutaka and Okinawa 100, and a breeder's clone Kanto 63, were involved in the present work. Originals and mutant clones which were isolated by various mutagenic treatments with gamma rays, EI and EMS were planted in an ordinary experimental field. A randomized block with three replications was employed. A plot had 20 plants in two rows. Five plants were taken for investigation from each plot. The plant density was 5555 plants per one-tenth of a hectare with a 0.6 m row and 0.3 m plant interval. The fertilizer level was 2.4−8.0−8.0 kg per one-tenth of a hectare as N−P−K.

Length of main stem, tuber weight per plant and number of tubers per plant were investigated immediately after the harvest. Small tubers of less than 50 g were discarded. Mutagenic treatments applied to induce the present mutant clones were as follows: gamma-ray irradiation, dosages 10−30 kR, 120 R/h dose rate; EI, 0.5% aqueous solution; EMS, 2 or 5% aqueous suspension. All the treatments were applied to the cuttings.

6.2. Results and discussion

Mutant clones which showed comparative changes in stem length from the original are shown in Table XII together with the other characteristics. In these clones it is observed that the other characteristics such as tuber yield, dry matter content and total sugar content had changed to some extent. Although short-vined mutants were induced only by gamma rays in both the cultivars Norin 1 and Tamayutaka, they were induced by chemical mutagen in Okinawa 100. There were no mutagenic differences in induction of dwarf mutation. Short vines induced by mutagenic treatment were characterized by the reduction of node number and not by internode length. As is clear from Table XII, short-vined mutant clones had decreased tuber weight. A significant correlation of r=0.578** between tuber weight and stem length was found in 23 clones and the originals. On the other hand, in Tachi-Norin 1, which is a spontaneous dwarf mutant of erect type [30], several mutants with long vines were induced by mutagenic treatment, though it is not clear whether by back-mutation or by the rearrangement of chimaerical structure. Correlation between tuber weight and stem length was also significant, being r=0.828*. Mutant clones with a change towards longer stem length increased tuber weight more than the original. From these facts, it is considered that the shortened vine may cause a reduction of leaf number and leaf area index (LAI) at the beginning of growth in the field, and may need more time to reach optimum LAI for tuber production. A short-vined mutant with larger LAI must be looked for as a first step in this respect. Mutant clones 0−100−GM−2 and K63−GM−9 are such a plant type though as yet insufficient.

In Table XII, mutant clones with changes in dry matter content and total sugar content in the tuber are shown. Some clones had extremely high content in both characteristics. N1—GM—9 was 20% higher than the originals in dry matter content and N1—GM—3 and TAM—GM—18 were approximately 50% higher than the originals in total sugar content. All these clones, however, had reduced tuber yield compared with the originals except N1—GM—3. The mutant clone N1—GM—3 is a promising clone having higher dry matter content and total sugar content without any reduction in tuber weight.

It is concluded that the mutations showed positive tendencies in agronomic characteristics as well in characteristics controlled by major genes.

7.   EFFECTS OF GAMMA IRRADIATION UPON INCOMPATIBILITY OF *Ipomoea leucantha* Jacq.

It has been reported by many workers that self-incompatibility and/or cross-incompatibility in certain plant species could be destroyed by use of ionizing radiation [31—33] and heat shock [34]. Sweet potato is a cross-pollinated plant and consists of variety groups in which a reproductive barrier exists. For this reason the conventional cross breeding of sweet potato has being carried out only by combinations of given pairs of compatible varieties. If techniques to break down this barrier were available, efficacy of cross breeding would greatly increase. Many attempts have been made to overcome this problem, but none have been successful [35].

We attempted in 1973 to break down the incompatibility of sweet potato by applying chronic irradiation by gamma rays to either pollen or style, but this definitely failed. The failure of the attempt is presumably due to the polyploidy of the sweet potato ($2n=90$, $6X$), being the gamete $3X$, and to exceedingly low frequency of phenotypic expression of mutations regarding the incompatibility gene action. It was, however, concluded that the incompatibility in sweet potato was controlled by a sporophytic reaction since it showed unilateral incompatibility and complete inhibition of microspore germination on the stigma unless there was a compatible combination.

It seemed wiser to use some diploid relatives which were controlled by a similar incompatibility system to the sweet potato. *Ipomoea leucantha* Jacq., one of the presumed prototypes of sweet potato, was employed for the experiment because of its diploidy and ease of flowering though it is non-tuberous.

7.1. Materials and methods

Ipomoea leucantha clones indigenous in Latin America were obtained through the Kyushu Agricultural Experiment Station, and were identified for their incompatibility relationships between clones. Flowering was induced by short day conditions (10 h day length) in two different conditions of day-time temperature.

Two separate irradiation experiments were conducted in the gamma phytotron equipped with a ^{137}Cs source at the Institute of Radiation Breeding.

7.1.1. Experiment A

Clones of *I. leucantha* were chronically irradiated with a dose rate of either 75 or 150 R/day from the beginning of lower bud formation to the flowering stage for about 1 month in two different conditions, 20°C and 30°C.

7.1.2. Experiment B

Clones were irradiated either for 20 h before pollination or 5 h after pollination with three different acute dosages of 5, 25 and 50 kR.

Pollination in the experiments was carried out in the laboratory with decapitated flowers which were castrated 1 day earlier. Compatibility was tested by microscopic observation on pollinated stigmas stained by cotton blue.

7.1.3. Experiment C

A crossing experiment was carried out to investigate seed set in incompatible combinations of the clones on the lines of Experiment A. The period of irradiation, however, was extended to two months before pollination.

7.1.4. Experiment D

A mentor pollen experiment was also carried out to find out the effect of pollen which was compatible but treated by a heavy dose of gamma rays (60 kR) so as to eradicate the ability to fertilize.

7.2. Results and discussion

From Experiment A, as shown in Table XIII, the effect of pollen chronically irradiated with gamma rays on cross-incompatibility was evident especially at the 30°C temperature, whereas it was not clear regarding self-incompatibility at both the irradiation dose rates. The depressive effect of gamma rays was no longer observed in compatible cross combinations. It is considered that some changes of incompatibility took place during the developmental stage of the floral organs and there was a phenotypic expression of high frequency only in haploid microspores rather than in somatic tissues (diploid) of the stigma. Increased effects of

TABLE XIII. EXPERIMENT A: EFFECT ON STIGMA IN DECAPITATED FLOWERS OF CHRONIC IRRADIATION BY GAMMA RAYS ON POLLEN GERMINATION

Cross combination	♀×♂ Dose rate (R/day)	Temperature condition (°C)	No. of pollinated flowers	No. of flowers with germinated pollen (%)	No. of germinated pollen	Mean No. of germinated pollen per stigma
Selfing of self incompatible clones	Cont.	Cont.	20	10	0 (0)	0	0
	"	"	30	118	1 (0.8)	3	0
	75	75	20	80	5 (6.3)	98	1.2
	"	"	30	141	1 (0.7)	1	0
	150	150	20	70	6 (8.6)	25	0.4
	"	"	30	115	0 (0)	0	0
Crossing between cross incompatible clones	Cont.	Cont.	20	15	0 (0)	0	0
	"	"	30	42	2 (4.8)	7	0.1
	"	75	20	15	1 (6.7)	2	0.1
	"	"	30	15	7 (46.7)	27	1.8
	"	150	20	15	1 (6.7)	3	0.2
	"	"	30	15	6 (40.0)	23	1.5
	75	Cont.	20	15	0 (0)	0	0
	"	"	30	15	0 (0)	0	0
	150	"	20	15	0 (0)	0	0
	"	"	30	15	0 (0)	0	0
Crossing between cross compatible clones	Cont.	Cont.	30	43	43 (100)	3302	76.8
	"	75	20	4	4 (100)	276	69.0
	"	"	30	5	5 (100)	289	57.8
	"	150	20	5	5 (100)	225	45.0
	"	"	30	5	5 (100)	320	64.0
	75	Cont.	20	15	15 (100)	1002	66.8
	"	"	30	15	15 (100)	938	62.5
	150	"	20	15	15 (100)	1026	68.4
	"	"	30	15	15 (100)	931	62.1

TABLE XIV. EXPERIMENT B: EFFECT ON STIGMA IN
DECAPITATED FLOWERS OF ACUTE IRRADIATION BY
GAMMA RAYS ON POLLEN GERMINATION

Cross combination	♀ × ♂ Dose (kR)		No. of pollinated flowers	No. of flowers with germinated pollen	Mean No. of germinated pollen per stigma
Irradiated for 20h before pollination					
Selfing of self-incompatible clone	Cont.	Cont.	15	0	0
	"	5	6	0	0
	"	25	6	0	0
	"	50	6	0	0
	5	Cont.	5	0	0
	25	"	6	0	0
	50	"	6	0	0
	5	5	8	0	0
	25	25	6	0	0
	50	50	8	1	0.1
Irradiated for 6h after pollination					
Selfing of self-incompatible clone	Cont.	Cont.	16	0	0
	5	5	18	1	0.1
	25	25	18	0	0
	50	50	18	0	0
Crossing between cross-compatible clones	Cont.	Cont.	12	12	155.8
	5	5	12	12	155.8
	25	25	12	12	169.7
	50	50	12	12	173.3

gamma rays at 30°C rather than at 20°C would be explained by the following three
reasons: (a) Difference of developmental speed of flowers due to temperature
conditions; flowers at 30°C were exposed to gamma rays from an earlier stage than
those at 20°C; (b) Induction of pseudo-compatibility by high temperature;
(c) Shift of radiation effect by the temperature condition.

Acute irradiation in Experiment B on floral organs before or after pollination
hardly affected incompatibility, as seen in Table XIV.

TABLE XV. EXPERIMENT C: EFFECT ON FRUIT AND SEED SET OF
CHRONIC IRRADIATION BY GAMMA RAYS ON POLLEN PARENT

Cross combination	Temperature condition (°C)	Dose rate (R/day)	No. of pollinated flowers (A)	No. of fruit (B)	Rate of fruit set(%) (C)*	No. of seed (D)	Rate of seed set(%) (E)**
Crossing between cross-compatible clones (L5-1 x L1-1)	24	Cont.	53	45	84.9	133	62.7
	"	125	69	55	79.7	128	46.4
	"	250	99	40	40.4	107	27.0
	32	Cont.	62	54	87.1	177	71.4
	"	125	55	48	87.3	140	63.6
	"	250	76	52	68.4	155	51.3
Crossing between cross-incompatible clones (L1-1 x L2-6)	24	Cont.	75	4	5.3	5	1.7
	"	125	125	85	68.0	314	62.8
	"	250	82	45	54.9	126	38.4
	32	Cont.	63	4	6.3	10	4.0
	"	125	67	6	9.0	11	4.1
	"	250	70	7	10.0	10	3.6
Selfing of self-incompatible clone (L1-1)	24	Cont.	85	11	12.9	20	5.9
	"	125	95	16	16.8	32	8.4
	"	250	70	9	12.9	14	5.0
	32	Cont.	84	8	9.5	13	3.9
	"	125	94	19	20.2	37	9.8
	"	250	83	21	25.3	26	7.8

* $C = \dfrac{B}{A} \times 100$

** $E = \dfrac{B}{4A} \times 100$; Each fruit contains four seeds when fertilization is complete.

   Results of Experiment C are shown in Table XV. In the compatible cross
combinations, rate of seed set decreased with increase in dose rate and with a
decrease in temperature. This is due to radiation hazard and temperature effect
on seed development. Breakage of self- and cross-incompatibility is obvious in
the selfing of clone L1-1 and in the cross combination of clones L1-1 × L2-6
respectively. A favourable dose rate for the breakage was 125 R/day in both cases
so far. Reaction to temperature of incompatibility did not agree with the results
obtained in Experiment A. The reason for this inconsistency seems to be the

TABLE XVI. EXPERIMENT D: EFFECT OF MENTOR POLLEN KILLED BY GAMMA RAYS ON FRUIT AND SEED SET

Cross combination		No. of pollinated flowers (A)	No. of fruit (B)	Rate of fruit set(%) (C)*	No. of seed (D)	Rate of seed set(%) (E)**
Crossing between cross-compatible clones	L5-1(Cont.)xL1-1(Cont.)	53	45	84.9	133	62.7
	" x L1-1(60kR)	48	3	6.3	4	2.1
	" x [L5-1(60kR)+L5-1(Cont)]	50	38	76.0	140	70.0
Crossing between cross-incompatible clones	L1-1(Cont.)x[L5-1(60kR) +L2-6(Cont.)]	50	1	2.0	1	0.5
Selfing of self-incompatible clones	L1-1(Cont.)xL1-1(Cont.)	85	11	12.9	20	5.9
	" x[L5-1(60kR)+L1-1(Cont)]	48	7	14.6	11	5.7
	L5-1(Cont)x[L1-1(60kR) +L5-1(Cont)]	49	1	2.0	1	0.5
	" x[L5-1(60kR) +L5-1(Cont)]	50	2	4.0	8	4.0

* $C = \dfrac{B}{A} \times 100$.

** $E = \dfrac{B}{4A} \times 100$; Each fruit contains four seeds when fertilization is complete.

difference in the irradiation period. In other words, the changes of incompatibility may be expressed at different times in relation to the temperature condition, since the length of developmental stage from flower bud initiation to maturation of gametophytes depends on the growth condition. It need more extensive studies to explain the interdependent relationship between irradiation effects and development conditions in this plant.

We expected that if the incompatibility system in *Ipomoea* had a relation to the antigen-antibody reaction, there would be effects from the mentor pollen method. Pollen irradiated acutely by 60 kR of gamma rays would lose its fertilizing ability without any changes in macro structures of proteins which might have a function on their receptors. However, no effects of the mentor pollen on the incompatibility system were observed in Experiment D (Table XVI). Some seeds were occasionally obtained by the method at a very low frequency in the cross-incompatible combination. This might be explained by parthenogenesis.

8. CONCLUSIONS

Mutagenic treatment with gamma rays, EI and EMS can induce changes in visible characteristics controlled by major genes in sweet potato. White tuber skin and colourless flesh of tuber are desirous in sweet potato cultivars for starch production. Mutants in such characters can be easily obtained by any mutagenic treatment. Short-vined mutants are also good material for practical breeding. It is needless to say that direct use of mutants is most promising in sweet potato. As for quantitative characters, variation can be induced by mutagenic treatment on clonal organs. Furthermore, the treatment on hybrid seed can increase variations in the segregating family.

In conventional sweet potato breeding in Japan, it can be postulated that the chance of any seedling becoming a beneficial cultivar is 1 in 100 000 to 150 000. In order to have a chance of being successful in mutation breeding, large amounts of material must be raised and a proper selection method must be used for superior types, because the genetic diversity is more limited in clonal progenies after mutagenic treatment than in families from hybrids. The target for mutation breeding should be highly concentrated on the particular point. Application of mutagens on the hybrid seed is recommended since it adds some genetic diversity by mutations to the original diversity of gene recombinations.

The answer to the questions what mutagen is most beneficial and what treatment is most efficient in practical mutation breeding in sweet potato is still not clarified. So far as the present experiments are concerned, gamma rays tended to show more drastic mutations and higher mutation frequencies than the other mutagens, EI and EMS.

Use of the adventitious bud technique in sweet potato with mutagenic treatment may be very significant, as Broertjes and van Harten pointed out [36]. Adventitious buds easily develop from leaf stalk, stem and root, and in vitro culture of these organs is simple and easy. Breakage of incompatibility must be accomplished in cultivated hexaploid sweet potato. These are important aspects which must be exploited in the future.

APPENDIX

Significance test of genetic variance by strain mean of plot unit is given as follows:

Block	r-1		
Derivative strain	v-1	M_1	$\sigma_e^2 + r\sigma_g^2$
Error	(r-1)(v-1)	M_2	σ_e^2

$$F = (M_1 \text{ of mutagenic treatment plot})/(M_1 \text{ of the control})$$

It involves genetic variance as well as any other sources of error.

Genetic variance = $(M_1 - M_2)/r$

$= \sigma_g^2$ . . . . .Total genetic variance among progenies within blocks.

Heritability in a general sense = σ_g^2/σ_P^2

Phenotypic variance = $\sigma_g^2 + \sigma_e^2$

$= \sigma_P^2$

When mean unit not plot unit,

$$\sigma_P^2 = \sigma_g^2 + \sigma_e^2/r$$

REFERENCES

[1] FUJISE, K., Spontaneous mutation and its use in sweet potato, Gamma Field Symp. 4 (1965) 43.

[2] MILLER, J.C., Further studies of mutations of the Port Rico sweet potato, Proc. Am. Soc. Hort. Sci. 33 (1935) 460.

[3] MASHIMA, I., SATO, H., X-ray induced mutations in sweet potato, Jpn. J. Breed. 8 (1959) 233.

[4] HERNANDEZ, T.P., HERNANDEZ, T., MILLER, J.C., Frequency of somatic mutations in sweet potato varieties, Proc. Am. Soc. Hort. Sci. 85 (1964) 430.

[5] SAKAI, K., Induction of mutants in sweet potato by absorption of ^{32}P, Gamma Field Symp. 5 (1966) 25.

[6] LOVE, J.E., "Mutation induction in the sweet potato by means of fast-neutron irradiation", Induced Mutations in Plants (Proc. Symp. Pullman, 1969) IAEA, Vienna (1969) 331.

232   KUKIMURA and KOUYAMA

[7] KUKIMURA, H., On the artificial induction of skin colour mutation of sweet potato (*Ipomoea batatas* [L.] Lam.) tuber, Technical News No.8, Inst. of Radiat. Breed. (1971).

[8] TAKEMATA, T., KUKIMURA, H., Studies on high sugar content in sweet potato breeding. 3. Variation of total sugars content and other traits in vM_2, vM_3 and F_1 families with mutagenic treatment, Jpn. J. Breed. **23** Suppl.2 (1973) 143 (Abstract in Japanese).

[9] KUKIMURA, H., TAKEMATA, T., Induced quantitative variation by gamma-rays and and ethylese-imine in tuber bearing plants, Gamma Field Symp. **14** (1975) 25.

[10] KUKIMURA, H., Effects of the recurrent treatment of mutagens on sweet potato plant, Jpn. J. Breed. **27** Suppl.2 (1977) 30 (Abstract in Japanese).

[11] MARUMINE, S., SAKAI, K., Studies on the induction of artificial induction in sweet potato. 1. X-ray induced mutations, Proc. Kyushu Crop Sci. Soc. **16** (1961) 4 (in Japanese).

[12] GUSTAFSSON, Å., GADD, I., Mutations and crop improvement. III. *Ipomoea batatas* [L.] Poir. (*Convolvulaceae*), Hereditas **53** (1965) 77.

[13] KUKIMURA, H., Mutation breeding in sweet potato, Nogyo gijutsu Kenkyusho Hokoku **33** (1978) 202 (in Japanese).

[14] CLOWES, F.A.L., Root apical meristems of *Fagus sylvatica*, New Phytol. **50** (1950) 1.

[15] ISHIHARA, K., Zonal structure of root apices, Gamma Field Symp. **2** (1963) 13.

[16] NISHIYAMA, K., MIYAZAKI, T., SAKAMOTO, S., Evolutionary autoploidy in the sweet potato (*Ipomoea batatas* (L.) Lam.) and its progenitors, Euphytica **24** (1975) 197.

[17] YEN, D.E., The Sweet Potato and Oceania, Bishop Museum Press, Honolulu (1974).

[18] KUKIMURA, H., Intervarietal difference of total sugars content in sweet potato tuber, Shiken Seisekisho (Annual Rep. Inst. of Radiat. Breed., 1972) (in Japanese).

[19] KUKIMURA, H., Effects of gamma-rays on segregation ratios in potato families, Potato Res. **15** (1972) 106.

[20] TAKAGI, Y., Studies on varietal differences of radiosensitivity in soybean, Bull. Inst. Radiat. Breed. No.3 (1974) 45 (in Japanese with English summary).

[21] RAWLINGS, J.O., HANWAY, D.G., GARDNER, C.O., Variation in quantitative characters of soybeans after seed irradiation, Agron. J. **50** (1960) 524.

[22] GREGORY, W.C., "Efficacy of mutation breeding", Mutation and Plant Breeding, Washington, NAS/NRC (1961) 461.

[23] MATSUO, T., ONOZAWA, Y., SHIOMI, M., Studies on mutations induced by radiations and chemicals: III. Variations of quantitative characteristics in vM_6 generation, Jpn. J. Breed. **14** (1964) 33 (in Japanese with English summary).

[24] RAJPUT, M.A., Increased variability in the M_2 of gamma-irradiated mung beans (*Phaseolus aureus* Roxb.), Radiat. Bot. **14** (1974) 85.

[25] HIRAIWA, S., TANAKA, S., "Effects of successive irradiation and mass screening for seed size, density and protein content of soybean", Seed Protein Improvement by Nuclear Techniques (Proc. Research Co-ordination Meetings, 1977) IAEA, Vienna (1978) 265.

[26] MARUMINE, S., SAKAMOTO, S., Characters and progeny test of mutated strains in sweet potatoes, Jpn. J. Breed. **27** Suppl.2 (1977) 28 (in Japanese).

[27] SAKAI, K.I., NISHIDA, T., OHBA, K., Statistical-genetic study on radiation-induced mutation in *Citrus*, Jpn. J. Breed. **38** Suppl.2 (1968) 25 (in Japanese).

[28] KUKIMURA, H., Effects of the recurrent treatment of mutagens on sweet potato plant, Jpn. J. Breed. **27** Suppl.2 (1977) 30 (in Japanese).

[29] DIVISION OF STATISTICAL INFORMATION, M.A.F.F., Statistics on Agriculture, Forestry and Fishery in Japan, 1979, Ministry of Agriculture, Forestry and Fishery, Tokyo (1979) (in Japanese).

[30] KYUSHU AGRICULTURE EXPERIMENT STATION, Characteristics of varieties and strains in sweet potato, Kyushu Noshi Kenkyu Shiryo No.43 (1972) (in Japanese).

[31] LEWIS, D., Structure of the incompatibility gene. III. Types of spontaneous and induced mutation, Heredity **5** (1951) 339.

[32] PANDEY, K.K., Elements of the S-gene complex. VI. Mutations of new self-incompatibility alleles, Genetica **41** (1970) 477.

[33] DE NETTANCOURT, D., ECOCHARD, E., Effects of chronic irradiation upon a self incompatible clone of *Lycopersicum peruvianum*, Theor. Appl. Genet. **38** (1968) 289.

[34] LINSKENS, H.F., "Biochemistry of incompatbility", Proc. 11th Int. Congr. Genet. **3** (1965) 629.

[35] FUJISE, K., Studies on flowering, seed setting and self- and cross-incompatibility in the varieties of sweet potato, Bull. Kyushu Agr. Exp. Sta. **IX** (1964) 123 (in Japanese with English summary).

[36] BROERTJES, C., van HARTEN, A.M., Application of Mutation Breeding Methods in the Improvement of Vegetatively Propagated Crops, Elsevier, Amsterdam (1978).

TETRAPLOID INDUCTION BY
GAMMA-RAY IRRADIATION IN MULBERRY*

K. KATAGIRI
Chubu Branch,
Sericultural Experiment Station, MAFF,
Matsumoto-shi, Nagano-ken

K. NAKAJIMA
Sericultural Experiment Station, MAFF,
Ohwashi, Yatabe-machi,
Tsukuba-gun, Ibaraki-ken,
Japan

Abstract

TETRAPLOID INDUCTION BY GAMMA-RAY IRRADIATION IN MULBERRY.
   Vigorously growing mulberry trees were exposed to 5 kR of gamma rays at the rate of
0.2 kR/h and 5 kR/h and successively pruned three times in two growing seasons. The frequency
of tetraploids induced was much higher than that of mutations, though almost all of them
were cytochimeras. By tracing a process of the formation of cytochimeras it is inferred that a
mutation is a unicellular event, with radiation treatment on materials in a multicellular constitution such as shoot apices resulting in the formation of chimeras, periclinal and mericlinal
chimeras.

Tetraploid induction and a process of the formation of cytochimeras following short-term chronic gamma-ray irradiation in mulberry are described.

MATERIALS AND METHODS

Experiment 1

One-year-old grafted mulberry trees of the variety Ichinose were used in
this experiment. The trees were planted in 1/2000-are pots, one per pot, at the
end of May and cut at 20 cm above the base just after planting. The materials
were grown in outdoor conditions. The uppermost bud on each tree was allowed
to grow by removing the others shortly after sprouting.

---

* Research carried out in co-operation with the IAEA under Research Agreement No. 2027.

TABLE I. FREQUENCIES OF MUTATIONS AND CYTOCHIMERAS APPEARING IN THE SECONDARY AND TERTIARY SHOOTS FOLLOWING GAMMA IRRADIATION OF MULBERRY

Figures in parentheses are percentages

Irradiation	Treatment[a]	No. of plants	Secondary shoot			Tertiary shoot			Newly appeared mutations and cytochimeras in tertiary shoot	
			No. of shoots	No. of mutated shoots	No. of cytochimeral shoots	No. of shoots	No. of mutated shoots	No. of cytochimeral shoots	No. of mutated shoots	No. of cytochimeral shoots
Low dose rate	A	20	80	2 (2.5)	4 (5.0)	224	0	11 (4.9)	0	7 (3.1)
	B	20	73	0	1 (1.3)	198	0	17 (8.5)	0	16 (8.0)
	C	20	76	0	0	178	1 (0.5)	8 (4.4)	1 (0.5)	8 (4.4)
High dose rate	A	12	44	0	2 (4.5)	144	0	6 (4.1)	0	0
	B	12	49	0	3 (6.1)	161	0	12 (7.4)	0	2 (1.2)
	C	13	66	0	1 (1.5)	156	1 (0.6)	9 (5.7)	1 (0.6)	6 (3.8)

[a] Treatment A, cutting the main shoot back to the 10th leaf above the beginning of the leafless portion; treatment B, cutting the main shoot back to the 5th leaf above the beginning of the leafless portion; treatment C, cutting the main shoot back to the beginning of the leafless portion.

About 20 leaves had expanded at the end of June when radiation treatment was conducted. The shoot tips were exposed to 5 kR of gamma rays from a ^{60}Co source at the rate of 0.2 kR/h (low dose-rate irradiation) and 5 kR/h (high dose-rate irradiation).

In the middle of August, after approximately another 30 leaves had expanded on a control plant, the irradiated shoot usually could be divided into three portions, the lower portion consisting of malformed narrow leaves, the middle portion of scaly leaves only (which was defined as a leafless portion) and the upper portion of normal leaves. The plants were randomly separated into three groups and the cutting-back treatment was conducted. That is, the shoots in the first group were pruned at the 10th leaf above the beginning of the leafless portion (treatment A), the shoots in the second group were pruned at the 5th leaf above the beginning of the leafless portion (treatment B) and the shoots in the third group were pruned at just above the beginning of the leafless portion (treatment C).

After the cutting-back treatment, secondary shoots from the upper five axillary buds were allowed to grow. In the following spring, just before bud-sprouting, the secondary shoots including both mutated and cytochimeral ones were pruned back to the 3rd leaf bud, and at the middle of June the remaining secondary shoots were pruned back again to below the 1st leaf bud. Observations on mutation and tetraploid were done on secondary and tertiary shoots developed on irradiated and unirradiated plants. The shoots which bore darker, thicker and broader leaves were tentatively selected as tetraploids according to Tojyo [1], and shoot tips and younger leaf tips were collected from them. The shoot tips were fixed in FAA for 24 h, dehydrated through the tertiary butyl alcohol series, embedded in Paraplast, cut in longitudinal 10 μm sections and stained in Delafield's haematoxylon. The sizes of nuclei and cells in shoot apices were examined on the prepared slides. Feulgen's squash method was applied to determine the chromosome number on the younger leaf tips collected. Along with the collection of these samples, the print of the lower leaf surface was made using the Sump method for measurement of the stomata.

Experiment 2

Vigorously growing shoots developing on the one-year-old grafts of the variety Ichinose were used in the second experiment as in the first experiment.

Radiation treatment was applied to the shoot tips as the targets at the end of May when about three leaves had expanded on the plants. The exposure was 5 kR and the exposure rate was 5 kR/h. Shoot tips were collected just after irradiation, one to six days, day by day, eight to 20 days at 2-day intervals, and 25, 27 and 31 days after irradiation, from 10 shoots at each sampling date. Killing, fixation, dehydration, embedding, making sections and staining were

KATAGIRI and NAKAJIMA

TABLE II. RELATIONSHIPS BETWEEN DOSAGES OF GAMMA RAYS, CUTTING-BACK TREATMENT AND TYPE OF CYTOCHIMERAS

Irradiation	Shoot	Treatment	Kind of mutation								
			2-4-4-4	2-2-4-4	2-2-2-4	2-4-2-2	2-4-4-2	4-2-4-4	4-2-4-2	4-2-2-2	(4-4-4-4)
Low dose rate	Secondary	A	2		1						
		B		1							
		C									
	Tertiary	A	6	1				1			
		B	12			3				3	
		C	4	1		1			1	2	1
High dose rate	Secondary	A	2								
		B	2								
		C	1								
	Tertiary	A	4								
		B	12								
		C	8								

carried out according to the procedure described in the first experiment. The size of nuclei and the number of chromosomes were examined on the prepared slides.

Fifteen unirradiated and 81 irradiated plants were reserved for cutting-back treatment. At the middle of July, after approximately 20 leaves had expanded on a control, the cutting-back treatment on three portions of the shoots developed was applied according to the procedure described in the first experiment. Shoot tips which were collected just before cutting-back treatment and those which were collected from secondary shoots developed after cutting-back treatment in the middle of August, about 30 days after cutting-back treatment, were handled under the same schedule as those described above for microscopic observations.

RESULTS

Experiment 1 (mutations and tetraploids after cutting-back treatment)

A few mutations were produced on the secondary and the tertiary shoots' by radiation treatment (Table I). That is, 'variegated leaf' and 'elongated leaf' on secondary shoots in treatment A, 'marginally curled leaf' on a tertiary shoot in treatment C, following low dose-rate irradiation, and 'elongated leaf' on a tertiary shoot in treatment C after high dose-rate irradiation.

In contrast, a large number of tetraploid secondary shoots developed on plants following both types of radiation treatment, although almost all of them were periclinal chimeras (Table II and Fig.1). The frequencies of tetraploid secondary shoots following the low dose-rate irradiation were inversely related to the amount of cutting-back. However, after high dose-rate irradiation, the highest production of tetraploids occurred in secondary shoots developed after the irradiated main shoots were cut back to the 5th leaf above the beginning of the leafless portion.

After cutting-back treatment, the secondary shoots produced more tetraploids in tertiary shoots developing on the irradiated plants. The frequency of tetraploids first appearing in tertiary shoots was highest in treatment B and the frequencies in treatments A and C were half of those for low dose-rate irradiations. However, the highest frequency of tetraploid tertiary shoots which appeared on plants following high dose-rate irradiation was in the shoots developed on the portion below the leafless portion, treatment C, one-third of this value in treatment B, and no tetraploids appeared in tertiary shoots on plants in treatment A.

Almost all the tetraploids were cytochimeras having one to several tetraploid cell layers in the shoot apex (Table II). Only one wholly changed tetraploid shoot

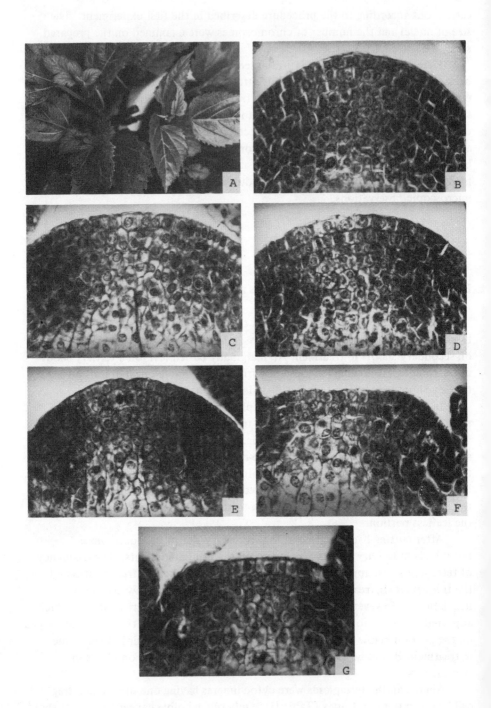

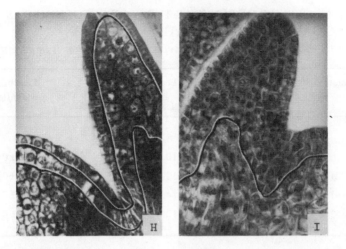

FIG.1. Cytochimeras induced by gamma-ray irradiation at the growing stage. A: Cytochimeral tertiary shoots appeared after cutting the secondary shoot back, cytochimeral tertiary shoots for two shoots on left and shoot of original Ichinose on right; B-G: Shoot apices; B: Original (2−2−2−2); C: Tetraploid (4−4−4−4); D: 4−2−2−2; E: 2−4−4−4; F: 2−2−4−4; G: 4−2−4−4; H: 2−4−2−2, diploid for epidermis, tetraploid for outer part and diploid for inner part of leaf primordium; I: 2−2−4−4, diploid for epidermis and outer part of leaf primordium, and tetraploid for inner part of leaf primordium.

was detected (Fig.1C). This tetraploid shoot developed on a secondary shoot derived from the main shoot which was cut back to just below the leafless portion, following low dose-rate irradiation. It was confirmed by microscopic observations on the shoot apices and the leaf surfaces that seven types of cytochimeras were induced in all tetraploid shoots. Seven kinds of cytochimeras occurred following low dose-rate irradiation, whereas only three types were observed in the shoots following high dose-rate irradiation. Among the cyto-chimeras the frequency of 2−4−4−4 (Fig.1E) was extremely high, followed by 4−2−2−2 (Fig.1D) and 2−4−2−2 (Fig.1H). A cytochimera of the 4−2−4−4 (Fig.1G) type was detected in tertiary shoots developed after cutting secondary shoots back to below the first leaf bud on the plant with treatment C, following low dose-rate irradiation. Larger stomata were visible on the leaf surface of a tetraploid (4−4−4−4) and a 4−2−2−2 chimera, whereas smaller stomata were scattered on the leaf surfaces of diploid (2−2−2−2) and other chimeras having diploid epidermises.

Microscopic observations on the leaf primordia of these cytochimeras showed that the epidermis and the internal tissues in 2−4−4−4 and 4−2−2−2 types were of the same ploidy levels as the cell layers in the apex. The youngest leaves of

TABLE III. LEAF SIZES OF SEVERAL TYPES OF CYTOCHIMERAS
FOLLOWING GAMMA (LOW DOSE-RATE) IRRADIATION
The number of shoots belonging to each type of cytochimera is in parentheses

Type of chimera	Leaf blade length	Leaf blade width	Leaf blade width/length
Original(2−2−2−2) (24)	14.4±1.7	10.9±1.3	0.75±0.03
2−4−4−4 (22)	14.4±2.2	12.7±2.3	0.88±0.06
2−4−2−2 (4)	15.2±1.4	14.0±2.4	0.91±0.07
4−2−2−2 (5)	15.3±0.9	12.6±1.1	0.81±0.04

2−4−2−2 shoots consisted of diploid inner tissues (Fig.1H). In 2−2−4−4 shoots only the inner part of the leaf primordium was tetraploid (Fig.1I), and no tetraploid tissue was observed in the leaf primordia of 2−2−2−4.

In general, tetraploid cytochimeras bore broader leaves compared with diploids (Table III). Among them, 4−2−2−2 shoots produced narrower leaves compared with 2−4−2−2 and 2−4−4−4 shoots. In these cytochimeras, the structures of axillary bud primordia which developed on the 3rd to 8th axillary bud sites, counted from the shoot apex downwards, were in the same ploidy arrangement as their shoot apices, except for 2−2−2−4 chimeras in which some axillary bud primordia became diploid.

Experiment 2 (process of formation of cytochimeras)

With the meristem proper more than three layers of cells were observed in each of the vegetative shoot apices, and the most distal three layers were well arranged to be designated as a tunica layer. Although the apical configuration and the cytohistological zonation patterns observed in the irradiated apices were distinctly different from those in the normal shoot tips, observations through a microscope were able to distinguish a single cell with a larger nucleus (Fig.2A) and cells having larger nuclei which appeared continuously in the first cell layer of the apex 10 days after irradiation (Fig.2B). At 12 days after irradiation masses of cells with larger nuclei were observed in approximately half of the apices irradiated (Fig.2C and 2D). Some of them included larger nuclei continuously in the middle parts of the second layers (Fig.2C), whereas in the

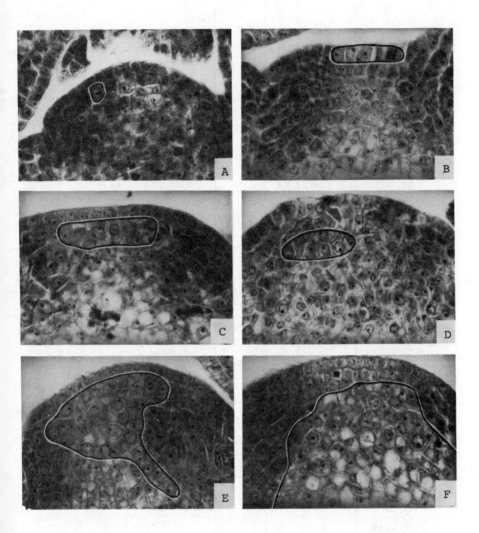

FIG.2.   Formation of cytochimeras following gamma-ray irradiation at the growing stage.
A: 8 days after irradiation, a cell with larger nucleus appeared in second cell layer; B: 10 days
after irradiation, cells with larger nuclei appeared continuously in first cell layer; C: 12 days
after irradiation, cells with larger nuclei appeared continuously in second and third cell layers;
D: 12 days after irradiation, cells with larger nuclei appeared continuously in third cell layer
and corpus; E: 20 days after irradiation, cells with larger nuclei appeared continuously from
second cell layer to inner part of corpus; F: 22 days after irradiation, cells with larger nuclei
appeared continuously from third cell layer to corpus.

TABLE IV. FREQUENCIES OF CYTOCHIMERAS APPEARING IN SECONDARY SHOOTS FOLLOWING GAMMA IRRADIATION AT THE THREE-LEAF STAGE IN MULBERRY

Figures in parentheses are percentages

Irradiation	Cutting-back treatment[a]	No. of plants	No. of secondary shoots	No. of cytochimeras		No. of cytochimeras originating from cytochimeral main shoots or cytochimeral secondary shoots	
				On plant	On secondary shoot	On plant	On secondary shoot
0 kR	B	15	50	0 (0)	0 (0)	—	—
5 kR at 5 kR/h	A	16	55	7 (43.7)	10 (18.1)	4 (57.1)	10 (100)
	B	16	52	7 (43.7)	8 (15.8)	3 (42.8)	3 (37.5)
	C	15	39	4 (26.6)	4 (25.0)	1 (25.0)	2 (50.0)

a   See footnote to Table I.

TABLE V. RELATIONSHIP BETWEEN CUTTING-BACK TREATMENT
AND TYPE AND FREQUENCY OF CYTOCHIMERAS FOLLOWING
GAMMA IRRADIATION AT THE THREE-LEAF STAGE

Cutting-back treatment	Type of cytochimera					Total
	2−4−4−4	2−2−4−4	2−4−2−2	2−4−4−2	4−2−2−2	
A	5	2	1	1	1	10
B	3	2	3			8
C	3				1	4

other apices cells with larger nuclei were allocated in the central parts of the
second and third layers (Fig.2D). The regions of cells with larger nuclei in the
apices were enlarged with the passing of time after irradiation (Fig.2E and 2F).

Thirty-three plants out of 81 plants observed were dead at the top of the
shoots irradiated. Therefore, 48 shoots were used for further treatment. Up to
the middle of July when cutting-back treatment was conducted, almost all
the shoots were forked. The relation of shoot cutting-back to frequencies of cytochimeras induced is shown in Table IV. The number of plants and shoots having
a cytochimeral constitution increased with increasing the height of cutting-back,
and the number of cytochimeral secondary shoots which originated from cytochimeral main shoots also increased with increasing the height of cutting-back.
Cytochimeras appearing after cutting-back treatment are shown in Table V. The
number of types of cytochimeras produced increased with increasing the height
of cutting-back.

DISCUSSION

Vigorously growing mulberry shoots were exposed to two types of short-
term chronic gamma rays to obtain information on improving the efficiency of
radiation treatment in mulberry to produce high frequencies of mutations [2].

The frequency of mutations induced in the present experiment was low in
comparison with that after dormant bud irradiation which produced 4.3% of
mutations in the secondary shoots on plants cut back to just above the leafless
portion [3]. Lapins and co-workers [4] reported that the mutation frequency
of summer buds of apple was low in comparison with that of the dormant
and the forced buds.

A high rate of tetraploid induction occurred after the cutting-back treatment. Fry [5] produced a comparatively large number of tetraploids by gamma irradiation in growing muscudine grapes, although all of them had a diploid epidermis over tetraploid internal tissues. Other instances of tetraploid induction by ionizing radiation have been reported for rice [6], tea [7], *Chrysanthemum morifolium* [8] and *Nicotiana tabacum* [9].

Following low dose-rate irradiation the highest frequency of tetraploid shoots was observed to occur in secondary shoots developed after cutting-back the main shoots to the 10th leaf above the beginning of the leafless portion, although the buds which developed these secondary shoots were apparently differentiated after irradiation. Therefore, the number of tetraploid shoots was in inverse relationship to the number of cells included in the apex at the time of irradiation. This means that the smaller number of cells located in the apex produced the larger cytochimeral sectors [10]. Many authors [4, 11 – 14] have also emphasized the importance of radiation treatment of a primordium when it consists of one or a few undifferentiated cells, in order to obtain as large a mutated sector as possible. The frequency of tetraploids in secondary shoots following high dose-rate irradiation was the highest in the group in which the irradiated main bud was cut back to the 5th leaf above the beginning of the leafless portion, followed by treatment A. This phenomenon may be explained as resulting from the killing of the centrally located cells in the apex by the irradiation. The same radiation effects may produce the reduction of cytochimeral tertiary shoots on the plants following high dose-rate irradiation. However, for low dose-rate irradiation, a comparatively large number of cytochimeral tertiary shoots developed as a result of the escape from radiation-killing of the cells which differentiated tertiary shoots. Bauer [15], Futsuhara [16], Nakajima [12] and Zwintscher [17] stressed the cutting-back of irradiated primary buds to induce high frequencies of mutations or to increase the rate of whole mutations, by forcing the shoots from the parts which were most likely to contain mutated cells. However, the results of the present experiments show that the deeper cutting-back will lead to the loss of many mutations which occur in the upper part of the shoot [3]. The apical meristem should be the centre of interest in inducing mutations, as mentioned by Lapins and co-workers [4].

Among the induced tetraploids, the frequency of 2—4—4—4 was over 50%. The reason why such a high rate of 2—4—4—4 shoots was produced is not clear, but it seems that this type of chimera is the most stable type, as suggested by the frequency of periclinal and anticlinal divisions in the second cell layer and inner tissues [18].

A difference of ploidy among the cell layers has been reported in many plants [19 – 25]. In the present experiment such histogenetic differences of cell layers were observed, which indicates that several primary layers in the apex of each shoot and axillary bud are histogenetically independent of one another [23]

A cell having a larger nucleus was observed in the shoot apex several days after radiation treatment, and a mass of cells with larger nuclei was observed a few weeks after radiation treatment. These phenomena mean that a mutation is a unicellular event [26].

In conclusion, as demonstrated in both experiments, growing shoots are more suitable targets for tetraploid induction than dormant buds, and since a mutation is a unicellular event radiation treatment on the materials with multi-cellular constitution such as shoot apices and dormant buds usually results in producing chimeras, periclinal and mericlinal chimeras.

ACKNOWLEDGEMENTS

The authors wish to thank T. Yokoyama and H. Hirano for their kind suggestions on conducting the experiments.

REFERENCES

[1] TOJYO, I., Studies on the polyploid in mulberry tree. I. Breeding of artificial auto-tetraploids, Bull. Sericult. Expt. Sta. 20 (1966) 187.
[2] NYBOM, N., "The use of induced mutations for the improvement of vegetatively propagated plants", Mutation and Plant Breeding, NAS-NRC Publ. 891 (1960) 252.
[3] KATAGIRI, K., Radiation damage in winter buds and relation of shoot cutting-back to mutation frequencies and spectra in acutely gamma-irradiated mulberry, Gamma Field Symp. 12 (1973) 63.
[4] LAPINS, K.O., BAILEY, C.H., HOUGH, L.F., Effects of gamma rays on apple and peach leaf buds at different stages of development. I. Survival, growth and mutation frequencies, Radiat. Bot. 9 (1969) 379.
[5] FRY, B.O., Production of tetraploid muscudine (V. rotundifolia) grapes by gamma radiation, Am. Soc. Hort. Sci. 83 (1963) 388.
[6] ICHIJIMA, K., On the artificially induced mutations and polyploid plants of rice occurring in subsequent generations, Proc. Imp. Acad. 10 (1934) 388.
[7] AMMA, S., Characteristics of tetraploid tea induced from gamma-irradiated Yabukita variety, Study of Tea 46 (1974) 1.
[8] ICHIKAWA, S., YAMAKAWA, K., SEKIGUCHI, F., TATSUNO, T., Variation in somatic chromosome number found in radiation-induced mutants of Chrysanthemum morifolium Hemsl. cv. Yellow Delaware and Delaware, Radiat. Bot. 10 (1970) 557.
[9] GOODSPEED, T.H., Occurrence of triploid and tetraploid individuals in X-ray progenies of Nicotiana tabacum, Univ. California Publ. Bot. 11 (1930) 299.
[10] SYBENGA, J., Quantitative analysis of radiation-induced sectorial discoloration of the leaf of Crotalaria intermedia, Radiat. Bot. 4 (1964) 127.
[11] GAUL, H., Present aspects of induced mutations in plant breeding, Euphytica 7 (1958) 275.
[12] NAKAJIMA, K., Induction of sports in roses by gamma-ray irradiation, Gamma Field Symp. 4 (1965) 55.

248   KATAGIRI and NAKAJIMA

[13] SPARROW, A.H., SPARROW, R.C., SCHAIRER, L.A., The use of X-rays to induce somatic mutations in *Saintpaulia*, Afr. Violet Mag. 13 (1960) 32.
[14] YAMAKAWA, K., SEKIGUCHI, F., Radiation-induced internal disbudding as a tool for enlarging mutated sectors, Gamma Field Symp. 7 (1968) 19.
[15] BAUER, R., The induction of vegetative mutations in *Ribes nigrum*, Hereditas 43 (1957) 323.
[16] FUTSUHARA, Y., Studies of radiation breeding in the tea plants, Gamma Field Symp. 6 (1967) 107.
[17] ZWINTSCHER, M., Die Auslösung von Mutationen als Methode der Obstzüchtung. I. Die Isolierung von Mutanten in Anlehnung an primäre Veränderungen. Züchter 25 (1955) 290.
[18] KATAGIRI, K., Radiation damage and induced tetraploidy in mulberry (*Morus alba* L.), Environ. Exp. Biol. 16 (1976) 119.
[19] DERMEN, H., The mechanism of colchicine-induced cytohistological changes in cranberry, Am. J. Bot. 32 (1945) 387.
[20] Periclinal cytochimeras and histogenesis in cranberry, Am. J. Bot. 34 (1947) 32.
[21] Periclinal cytochimeras and origin of tissues in stem and leaf of peach, Am. J. Bot. 40 (1953) 154.
[22] Pattern of tetraploidy in the flower and fruit of a cytochimeral apple, J. Hered. 44 (1953) 31.
[23] Nature of plant sports, Am. Hort. Mag. 39 (1960) 123.
[24] SATINA, S., BLAKESLEE, A.F., AVERY, A.G., Demonstration of the three germ layers in the shoot apex of *Datura* by means of induced polyploidy in periclinal chimeras, Am. J. Bot. 27 (1940) 895.
[25] SATINA, S., BLAKESLEE, A.F., Periclinal chimeras in *Datura stramonium* in relation to development of leaf and flower, Am. J. Bot. 28 (1941) 895.
[26] BROERTJES, C., "Production of polyploids by the adventitious bud technique", Polyploidy and Induced Mutations in Plant Breeding (Proc. Panel Bari, 1972), IAEA, Vienna (1974) 29.

STUDIES ON MUTATION BREEDING IN MULBERRY (*Morus* spp.)*

H. FUJITA
Institute of Radiation Breeding, NIAS, MAFF,
Ohmiya, Ibaraki-ken

M. WADA
Sericultural Experiment Station, MAFF,
Ojiya, Niigata,
Japan

Abstract

STUDIES ON MUTATION BREEDING IN MULBERRY (*Morus* spp.).
Re-irradiation of induced mulberry mutants with gamma rays has proved to give higher mutation frequencies and a wider mutation spectrum than when the original cultivars were irradiated. A comparison between chronic and acute re-irradiation was made, using a special cutting-back technique. Mutation frequencies of the shoots that developed from the sub-lateral shoot in chronic irradiation were lower than in acute irradiation. The five-lobed cultivar Ichinose raised an entire-leaved mutant, IRB240-1, by gamma rays, which showed reversion in leaf shape, from entire to lobed, by re-treatment of the mutant with gamma rays. The mutant and original cultivar were crossed with an entire-leaved cultivar Shiromekeiso, which was considered to be homozygous *(11)* as to entire leaf shape. As a result of these crosses, the three mutants are supposed to be not of genic origin concerning leaf shape. A drop of conidia suspension was placed on a scratched surface of the irradiated shoots to select resistant mutants to die-back disease. Two resistant strains were selected by means of inoculation of conidia. It is considered that inoculation of solution using a vaccination apparatus was most efficient and reliable for selecting resistant mutants. Thirteen mutants and three strains were tested for the rooting ability of semi-softwood cuttings. There were some mutant strains that did not show any disability of rooting initiation of the shoot. It is considered possible that useful mutants which show high rooting ability can be selected by gamma irradiation. In general, when the plant is exposed to gamma rays, it does not only change in visible character, but also in fine structure. To detect the change in fine structure of mulberry a scanning electron microscope was used. The mutants not only varied in visible character from the original, but also in invisible changes, such as trichome, idioblast, etc.

1. INTRODUCTION

Mulberry, *Morus* spp., is one of the vegetatively propagated species that has a wide natural distribution in sub-tropical regions and it is important because it is the only diet for silkworms.

---

* Research supported by the IAEA under Research Contract No. 1336.

249

Mutation research has been performed since 1957 on mulberry. We have isolated about fifty mutants and two of them have been adopted for local trials [1, 2]. Fujita and Nakajima [3] used roses as test plants to study the best time for irradiation and the best exposure and exposure rate, while paying special attention to cutting-back methods to increase the percentage of wholly mutated mutants. Using the cutting-back method, it was reported that re-irradiation provides an optimum chance for mutated cells in lateral and sub-lateral buds to increase the mutation frequency and mutation spectrum in mulberry [3]. But the problem remains, which is better for obtaining the mutant, acute re-irradiation or chronic re-irradiation?

The main aims of mulberry breeders are higher leaf yield, good leaf quality, resistance to disease, higher rooting ability of cuttings, etc.

The results of research carried out on five topics are described here.

2.   INDUCTION OF MUTATION

2.1. Comparison between chronic re-irradiation and acute re-irradiation

A gamma-field has been advocated as the most suitable and convenient radiation facility for whole-plant irradiation of vegetatively propagated woody perennials. At the Institute of Radiation Breeding gamma-field various kinds of mutations have arisen in apple [4], roses [5], mulberry [6], etc., from whole-plant irradiation, acute as well as chronic. Little information, however, is available on successive exposures extending over five years.

In the experiment that was carried out to compare chronic and acute irradiation an important problem was encountered concerning the materials to be used. For example, it is doubtful that a bud on the current shoot actually was treated with 10 kR, although it developed from a plant which had received gamma rays of 10 kR for five successive years. Therefore, it is inappropriate to compare this bud with a bud which is exposed to acute irradiation of 10 kR for five years. The problem can be resolved by using a bud that has remained latent throughout the period of irradiation (in the case of chronic treatment) and a bud that is still latent after irradiation (in the case of acute treatment). The latent buds, in general, are located at the base of branches or shoots and can develop into shoots by cutting-back to the base. In addition, the cutting-back technique has been successively applied in vegetatively propagated woody perennials [5], including mulberry, because of the mutated shoots.

In some crops, re-treatment of induced mutants by gamma rays (conveniently termed 're-irradiation') has given rise to some interesting results concerning mutation frequencies and mutation spectra. A few examples of mulberry

have already been given elsewhere [3]. A comparison between chronic and
acute re-irradiation of mulberry is dealt with here.

2.1.1. Materials and methods

A mutant strain, IRB240-1, was used in the re-irradiation experiment.
The mutant strain originated from a mutant shoot appearing in a plant that
had been grown 41 m from the ^{60}Co source of the gamma-field for three years
under chronic irradiation, with an average daily exposure of 8.8 R. Besides
the changes in leaf shape, it showed an increase both in dry matter weight
per unit area and in the length of new shoots in spring, compared with the
original cultivar [7]; both changes are considered to be favourable for
silkworm raising.

In acute irradiation, one-year-old grafts of IRB240-1 and the original
Ichinose were irradiated in the gamma-room in late April 1975 with 6.5, 10,
15, 20 and 35 kR at various exposure rates (Table I). The previous year's
shoots of grafts were pruned 20–30 cm before irradiation (Fig.1(a)). The
underground part of the plants was shielded with lead blocks to avoid radiation
damage. The irradiated grafts were planted in a field with plants spaced at
50 cm and with a row space of 300 cm. In May, all sprouting buds except the
uppermost three or four buds were removed, and in June the surviving buds
and the axillary buds adjacent to the inactive buds were frequently observed
to sprout and to develop into shoots. These lateral shoots were allowed to
elongate without any cutting-back procedure to allow the plant to recover from
radiation damage. In the following early spring all the shoots were pruned,
10 buds remaining from the base of the shoots (Fig.1). The second cutting-back
was performed on the lateral shoots before sprouting in February – March of
the following year. The point to be cut-back on lateral shoots in the second
cutting-back treatment varied, depending on the position where a lateral
shoot was situated on the shoot [3].

In chronic re-irradiation IRB240-1, which had been grown in the gamma-
field for five years from 1970 to 1975, and Ichinose, which had been treated
for six years from 1969 to 1975, were used (Fig.1(b)). Both materials were
planted 34.5–87 m from the ^{60}Co source with plants spaced at 3.5 m.
Although exposure reduces exponentially as the plants are located farther
from the source, both the IRB240-1 and Ichinose plants were broadly
classified into five exposure groups (Table I). Exposure for acute re-irradiation
was established so as to correspond roughly to each of the five exposures for
chronic re-irradiation. Irradiated plants were transplanted from the gamma-field
to a field with a row space of 300 cm and plants spaced at 100 cm in late
March 1975. Immediately after transplanting, the plants were cut-back to
the base of the sub-lateral branches (1972 shoots for IRB240-1, 1971 shoots

TABLE I. MUTATION FREQUENCIES OF 1-YEAR-OLD GRAFTS OF THE INDUCED MULBERRY MUTANT STRAIN IRB240-1 AND ITS ORIGINAL CULTIVAR ICHINOSE AFTER GAMMA-RAY IRRADIATION (FIRST CUTTING-BACK)

Cultivar and mutant strain	Treatment[a]	Total exposure (kR)	Exposure rate	No. of treated plants	No. of survivals[b]	No. of shoots	No. of mutated shoots	No. of mutated shoots with leaf shape: 2,3-lobed	5-lobed	Many lobed	Entire	Lobate	Deformed
IRB240-1	Acute	35	0.37 kR/h	49	47(96)	141	5(3.6)		5(3.6)				
	Chronic	27.2–34.8	22.0–28.0 R/d	6	6(100)	139	1(0.7)	1(0.7)					
	Acute	20	0.37	47	43(91)	129	2(1.6)	1(0.8)	1(0.8)				
	Chronic	18.9–22.3	15.2–17.9	8	8(100)	346	14(4.0)		14(4.0)				
	Acute	15	0.37	46	26(57)	78	3(3.9)		1(1.3)			1(1.3)	1(1.3)
	Chronic	11.9–15.9	9.6–12.8	9	7(78)	115	2(1.7)						2(1.7)
	Acute	10	0.25	48	44(92)	132	1(0.8)		1(0.8)				
	Chronic	8.1–10.4	6.5–8.4	16	16(100)	360	0						
	Acute	6.5	0.17	50	49(98)	147	0						
	Chronic	5.7–7.3	4.6–5.9	26	26(100)	509	0						
		0	0	30	30	90	0						
Ichinose	Acute	35	0.37	72	68(94)	204	5(2.5)	2(1.0)			1(0.5)	1(0.5)	1(0.5)
	Chronic	27.9–43.5	15.3–28.3	10	10(100)	167	3(1.8)	3(1.8)					
	Acute	20	0.37	77	28(36)	84	9(10.8)	2(2.4)			6(7.1)		1(1.2)
	Chronic	20.0–23.7	9.7–12.9	8	8(100)	99	8(8.1)	4(4.0)		3(3.0)	1(1.0)		
	Acute	15	0.37	75	57(76)	171	13(7.6)	1(0.6)			11(6.4)	1(0.6)	
	Chronic	13.1–14.9	6.6–8.5	10	9(90)	156	1(0.6)						1(0.6)

Cultivar and mutant strain	Treat-ment[a]	Total exposure (kR)	Exposure rate	No. of treated plants	No. of survivals[b]	No. of shoots	No. of mutated shoots	No. of mutated shoots with leaf shape:					
								2,3-lobed	5-lobed	Many lobed	Entire	Lobate	Deformed
Ichinose (cont.)	Acute	10	0.25	76	65(86)	195	10(5.1)	3(1.5)			4(2.1)	1(0.5)	2(1.0)
	Chronic	8.2–11.5	5.3–5.9	23	23(100)	368	15(4.1)	6(1.6)			9(2.4)		
	Acute	6.5	0.17	51	48(94)	144	1(0.7)				1(0.7)		
	Chronic	5.5–7.2	3.5–4.7	28	28(100)	390	1(0.3)					1(0.3)	
		0	0	30	30	90	0						

[a] Acute: lateral shoot basis.

[b] Chronic: second sub-lateral shoot basis.
Figures in parentheses are percentages.

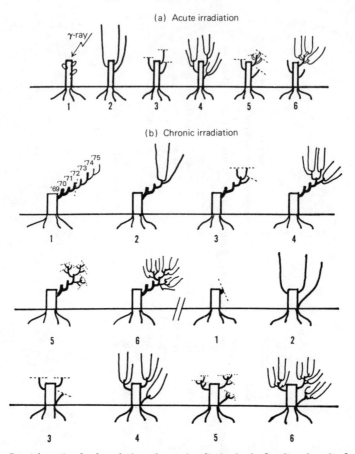

(a) Acute irradiation

(b) Chronic irradiation

FIG.1(a).   Special cutting-back technique (acute irradiation): 1. Irradiated graft; 2. Lateral shoots developed and elongated; 3. First cutting-back performed at tenth leaf of lateral shoots; 4. Sub-lateral shoots develop; 5. Second cutting-back, the point of cut-back varying, depending on the position where the sub-lateral shoot is situated on the shoot, e.g. one bud remains for the uppermost shoot; 6. Second sub-lateral shoots develop. The dotted line indicates the point of cut-back. (b) Cutting-back techniques (chronic irradiation): the first 1–6 represent continuous cutting-back and the second 1–6 represent more severe cutting-back (1969–1975).

for Ichinose). The second sub-lateral shoots that developed from the base were left to elongate. In early March of the following year the plants were cut-back to the base of the lateral shoots (1970 shoots for IRB240-1, 1969 shoots for Ichinose). The axillary buds that had remained latent for five or six years in the stem sprouted and elongated. These developed shoots were handled by a special cutting-back technique [3], in order to force latent buds developing into new shoots. Shoots developing from the lower part of the stem were cut-back above the tenth leaf in early July of the year of second cutting-back.

Observations of mutation were made on leaf shape, using lateral shoots in the case of acute re-irradiation. Mutation frequencies were scored on a shoots basis and percentages of wholly mutated shoots were estimated.

2.1.2. Results

When IRB240-1 and Ichinose were treated with acute re-irradiation, frequencies of wholly mutated shoots were higher than in the chronic treatment, except at 20 kR of Ichinose (Table I). Most of the mutations induced in IRB240-1 and Ichinose showed a change in leaf shape, such as 2,3-lobed and deformed. Twenty-two 5-lobed mutations from IRB240-1 showed an apparent resemblance in leaf shape to the original cultivar Ichinose and also 32 entire mutations resembled IRB240-1. Although it is somewhat difficult to form a conclusion because of the small number of plants, there was a tendency for the 5-lobed mutants to be mostly induced at 18.9—22 kR from chronic irradiation and at 35 kR from acute irradiation.

After the second cutting-back to the base of the lateral shoots, frequencies of wholly mutated shoots were higher than in the acute treatment and also higher than the frequency of shoots that developed from the 1972 branches (Table II). But when Ichinose was treated with chronic irradiation, mutation frequencies of wholly mutated shoots were lower than in the acute treatment, except at 35 kR and 10 kR. When the plants irradiated with chronic irradiation were given the second cutting-back treatment, lateral shoots that had developed from the upper part of the stump, compared with the shoots that had developed from the lower part of the stump, were more mutated (Table III). Sixty-nine 5-lobed mutations from IRB240-1 and 73 entire-leaved mutations from Ichinose were induced. In the first year's results, 18.9—22 kR chronic irradiation and 35 kR acute irradiation mostly induced 5-lobed mutation, but 5-lobed mutation was induced at over 15 kR with acute irradiation after the second cutting-back treatment. Entire-leaved mutation was induced with acute irradiation at exposures of 20, 15 and 10 kR. The total mutation frequencies of IRB240-1 and Ichinose were 4.9 and 5.5%, respectively. Both mutation frequencies were increases compared with the first year's results, 1.4 and 3.8%.

After the second cutting-back to the 1969 shoot of IRB240-1 and the 1970 shoot of Ichinose, two or three second sub-lateral shoots developed from each sub-lateral shoot, but many of the sub-lateral shoots that developed from the upper part of the stumps were reduced by damage of the cambium by long-horned beetles and did not produce second sub-lateral shoots. On the other hand, sub-lateral shoots that had developed from the lower part of the stump produced three or more second sub-lateral shoots (Tables IV and V).

TABLE II. MUTATION FREQUENCIES DEVELOPED FROM WHOLLY MUTATED SHOOTS OF THE INDUCED MULBERRY MUTANT STRAIN IRB240-1 AND ITS ORIGINAL CULTIVAR ICHINOSE AFTER GAMMA-RAY IRRADIATION (SECOND CUTTING-BACK) (total exposure, exposure rate and number of treated plants as in Table I)

Cultivar and mutant strain	Treatment	No. of survivals	No. of shoots	No. of mutated shoots[c]	No. of mutated shoots with the leaf shape:									
					2,3-lobed	5-lobed	Many lobed	Entire	Elongated	Deformed	Lobate	Erose	Mucronate	Other
IRB240-1	Acute	42	641[a]	44(6.9)	20	21			1				1	1
	Chronic	6	20[b]	6(30.0)	6									
	Acute	38	661	19(2.9)	9	7			1	1				
	Chronic	8	65	20(30.8)	6	14								
	Acute	22	406	39(9.6)	6	21				12				
	Chronic	9	64	12(18.8)	12									
	Acute	36	520	12(2.3)	5				1	2		4		
	Chronic	14	137	9(5.8)	8				1					
	Acute	49	406	16(3.9)	10	6								
	Chronic	24	175	28(16.0)	28									
	Chronic	10	76	0										
Ichinose	Acute	56	553	14(2.5)	1		6			2	2	1		2
	Chronic	10	77	16(20.8)			14				2			
	Acute	27	246	47(19.1)	5			39	1	2				
	Chronic	8	39	6(15.4)			4			1	1			
	Acute	57	690	62(8.9)	15		11	34		2				
	Chronic	10	46	1(2.2)			1							

Cultivar and mutant strain	Treatment	No. of survivals	No. of shoots	No. of mutated shoots[c]	No. of mutated shoots with the leaf shape:									
					2,3-lobed	5-lobed	Many lobed	Entire	Elongated	Deformed	Lobate	Erose	Mucronate	Other
Ichinose (cont.)	Acute	67	806	25(3.1)	2		4			17	1	1		
	Chronic	21	105	7(6.7)	1		3			1	2			
	Acute	48	586	18(3.1)	2		2		1	7	4			2
	Chronic	23	120	0										
		10	83	0										

a Second sub-lateral shoot basis.

b Lateral shoot basis.

c Figures in parentheses are percentages.

TABLE III. MUTATION FREQUENCIES OF THE UPPER AND LOWER
PARTS OF THE INDUCED MUTANT STRAIN IRB240-1 AND ITS
ORIGINAL CULTIVAR ICHINOSE AFTER CHRONIC GAMMA-RAY
IRRADIATION (SECOND CUTTING-BACK)

Cultivar and strain	Exposure (kR)	No. of shoots lower part of stump	No. of mutated shoots[a]	No. of shoots upper part of stump	No. of mutated shoots[a]
IRB240-1	27.2–34.8	37	2(5.4)	30	17(56.7)
Ichinose	27.9–43.5	6	2(33.3)	51	15(29.4)
IRB240-1	18.9–22.3	21	4(19.0)	32	5(15.6)
Ichinose	20.0–23.7	18	3(16.7)	8	3(37.5)
IRB240-1	11.9–15.9	31	6(19.4)	20	4(20.0)
Ichinose	13.1–14.9	22	0	20	1(0.5)
IRB240-1	8.1–10.4	113	14(12.4)	79	11(13.9)
Ichinose	8.2–11.5	45	0	60	6(10.0)
IRB240-1	5.7–7.3	38	10(26.3)	31	2(0.6)
Ichinose	5.5–7.2	53	0	74	1(1.4)
Total IRB240-1		240	36(15.0)	192	39(20.3)
Ichinose		144	5(3.5)	213	26(12.2)

[a] Figures in parentheses are percentages.

        In comparing acute and chronic irradiation, mutation frequency in
acute irradiation is higher than in chronic irradiation, although mutation
frequency of the first cutting-back treatment gave opposite results. The
reason for low frequencies has already been mentioned.

2.1.3. Discussion

        The cutting-back technique after transplanting from the gamma-field
enhanced the induction of mutation in irradiated plants treated with chronic
irradiation [8]. In this experiment, mutation frequencies of IRB240-1 and
Ichinose were higher than those which remained in the gamma-field. Most of
the mutations induced from the two strains showed a change in leaf shape.
Among them, 5-lobed mutants showed an apparent resemblance in leaf shape
to the original cultivar Ichinose. On the assumption that 5-lobed mutation is a
reversion to the original Ichinose, acute irradiation, compared with chronic
irradiation, needs a higher exposure to induce reverse mutation.

TABLE IV. MUTATION FREQUENCIES DEVELOPED FROM THE FIRST
SUB-LATERAL SHOOTS OF THE INDUCED MUTANT STRAIN IRB240-1
AND ITS ORIGINAL CULTIVAR ICHINOSE AFTER CHRONIC GAMMA-RAY
IRRADIATION (SECOND CUTTING-BACK)

Cultivar and strain	Exposure (kR)	No. of shoots lower part of stump	No. of mutated shoots[a]	No. of shoots upper part of stump	No. of mutated shoots[a]
IRB240-1	27.2−34.8	87	0	25	1(4.0)
Ichinose	27.9−43.5	54	6(11.1)	150	9(6.0)
IRB240-1	18.9−22.3	104	22(21.2)	51	4(7.8)
Ichinose	20.0−23.7	112	2(1.8)	105	6(5.7)
IRB240-1	11.9−15.9	72	2(2.8)	120	1(0.8)
Ichinose	13.1−14.9	137	4(2.9)	52	39(5.7)
IRB240-1	8.1−10.4	329	1(0.3)	61	0
Ichinose	8.2−11.5	303	6(2.0)	208	4(1.9)
IRB240-1	5.7−7.3	446	17(3.8)	179	1(0.6)
Ichinose	5.5−7.2	447	16(3.6)	58	8(13.8)
Total IRB240-1		1038	42(4.05)	436	7(1.61)
Ichinose		1053	34(3.23)	578	38(5.24)

[a]   Figures in parentheses are percentages.

The entire-leaved mutation was mostly at 8.2−11.5 kR with chronic
irradiation and at over 15 kR with acute irradiation. Special attention should
be paid to the former exposure, IRB240-1 being induced from Ichinose at
an exposure of 9.5 kR with chronic irradiation.

After the second cutting-back to the base of the 1969 shoots with chronic
irradiation, lateral shoots that had developed from the upper part of the stump,
compared with shoots that had developed from the lower part of the stump, were
more mutated. The reason why lateral shoots developed from the upper part
of the stump was considered to be the higher frequency of the wholly mutated
shoots rather than the lower one, the former being mutated with more exposure
than the latter because the latter were covered by earth during chronic irradiation
in the gamma-field.

The more the plants were exposed, the more the mutation frequencies
increased and usually acute irradiation induced mutation more frequently than
chronic irradiation, but after the second cutting-back treatment chronic irradiation
induced more mutations. Many reports concerning comparison of acute and

TABLE V. MUTATION FREQUENCIES DEVELOPED FROM SUB-LATERAL SHOOTS OF THE INDUCED MUTANT STRAIN IRB240-1 AND ITS ORIGINAL CULTIVAR ICHINOSE AFTER GAMMA-RAY IRRADIATION (MORE SEVERE CUTTING-BACK)

Cultivar and mutant strain	Treatment	Total exposure (kR)	Exposure rate	No. of treated plants	No. of survivals	No. of shoots	No. of mutated shoots[a]	No. of mutated shoots with the leaf shape of:									
								2,3-Lobed	5-Lobed	Many Lobed	Entire	Elongated	Deformed	Lobate	Erose	Mucronate	Other
IRB240-1	Acute	35	0.37 kR/h	49	42	641	44(6.9)	20	21			1			1	1	
	Chronic	27.2–34.8	22.0–28.0	6	4	112	1(1.0)	1									
	Acute	20	0.37	47	38	661	19(2.9)	9	7			1	1				1
	Chronic	18.9–22.3	15.2–17.9	8	8	155	26(16.8)	13	13								
	Acute	15	0.37	46	22	406	39(9.6)	6	21				12				
	Chronic	11.9–15.9	9.6–12.8	9	7	192	3(1.6)	2					1				
	Acute	10	0.37	48	36	520	12(2.3)	5				1	2		4		
	Chronic	8.1–10.4	6.5–8.4	16	14	390	1(0.3)	1									
	Acute	6.5	0.37	50	49	406	16(3.9)	10	6								
	Chronic	5.7–7.3	4.6–5.9	26	22	625	18(2.9)	16				3					
		0	0	30	10	76	0										
Ichinose	Acute	35	0.37	72	56	553	14(2.5)	1		6			2	2	1		2
	Chronic	27.9–43.5	15.3–28.3	10	9	204	15(7.4)	3		12				1			
	Acute	20	0.37	77	27	246	47(19.1)	5			39		2	1			
	Chronic	20.0–23.7	9.7–12.9	8	8	217	8(3.7)	1		6		1	1				
	Acute	15	0.37	75	57	690	62(8.9)	15		11	34		2				
	Chronic	13.1–14.9	6.6–8.5	10	9	189	7(3.7)	2		4							
	Acute	10	0.37	76	67	806	25(3.1)	2		4			17	1	1		1
	Chronic	8.2–11.5	5.3–5.9	23	19	511	10(2.0)	2		5	1		1	2			
	Acute	6.5	0.37	51	48	586	18(3.1)	2		2		1	7	4			1
	Chronic	5.5–7.2	3.5–4.7	28	23	505	24(4.8)	6		16			2				2
		0	0	30	10	83	0										

[a] Figures in parentheses are percentages.

chronic irradiation have suggested that acute irradiation induced more mutations than chronic irradiation [9]. There are few reports which support these results. Intense cutting-back treatments and re-irradiation by gamma rays might contribute to an increase in mutation frequencies in chronic re-irradiation.

3.   GENETICS AND SELECTION

3.1. Breeding behaviour in induced mutants of mulberry

In pear, carnations and chrysanthemums, some characters of spontaneous mutants reverted to the original ones when the mutants were treated with gamma rays [10–12]. The 5-lobed mulberry cultivar Ichinose raised an entire-leaved mutant, IRB240-1, by gamma rays. This mutant then showed reversion in leaf shape, from entire to lobed, by re-treatment of the mutant with gamma rays. The breeding behaviour of induced mutants is dealt with here to clarify the mechanism responsible for reversion.

3.1.1. Materials and methods

Materials used in the breeding test were the cultivar Ichinose and IRB240-1, the cultivar Kairyonezumigaeshi No. 3183 and an entire-leaved spontaneous mutant. The origins and characteristics of IRB240-1 have already been described, and No. 3183 is an entire-leaved mutant strain from 5-lobed Kairyonezumigaeshi. The mutant was induced by gamma-ray irradiation of one-year-old grafts with 5 kR at the rate of 5 kR/h. It is characterized by a change in leaf shape from 5-lobed to entire and, in addition, its internodal length shows a decrease of approximately 20% over the original cultivar. An entire-leaved spontaneous mutant of Kairyonezumigaeshi originated from a mutant shoot, which appeared in 1961 in a plant that had been grown in the field at the Tohoku Branch of the Sericultural Experiment Station.

An entire-leaved cultivar Shiromekeiso was used as a tester in the breeding test. The mutants and original cultivars were crossed with Shiromekeiso in May 1970. Seeds were sown in jiffy pots or beds filled with sterilized soil in July 1970, and the seedlings were grown in a greenhouse or in beds until April 1972, when they were transplanted in the field.

Observations were done on the leaf shape of the seedlings over 70 cm in height in the period from August – September 1972.

3.1.2. Results and conclusions

As can be seen in Table VI, the seedlings from these cross combinations segregate with lobed leaf and entire leaf and segregation ratios are almost

TABLE VI. SEGREGATION OF SEEDLINGS WITH ENTIRE LEAF AND LOBED LEAF AFTER CROSSES
BETWEEN MUTANTS OR ORIGINALS AND SHIROMEKEISO

Cross combination	Total No. of seedlings observed	Entire leaf	No. of seedlings with Lobed leaf	Both entire and lobed leaves
IRB240-1 × Shiromekeiso	68	42	26	0
Ichinose × Shiromekeiso	68	41	27	0
No. 3183 × Shiromekeiso	66	34	31	1
Entire-leaved spontaneous mutant × Shiromekeiso of Kairyonezumigaeshi	21	7	13	1
Kairyonezumigaeshi × Shiromekeiso	39	22	17	0

the same in all the crosses. There appear to be few seedlings with both entire and lobed leaves.

In mulberry, the lobed leaf is dominant over the entire leaf and the lobeness is determined by a single gene; it is suggested that the genotype of Kairyone-zumigaeshi is heterozygous *(L1)* because the selfed progenies show a segregation ratio of lobed to entire 3:1 [13]. The cultivar Shiromekeiso did not give rise to a lobed plant in the first and second progenies (M. Matsushima, personal communication) and therefore the tester is considered to be homozygous *(11)* as to entire leaf shape.

The sum of chi-square is 0.642 in the cross combination of Kairyonezumi-gaeshi X Shiromekeiso. It indicated that the observed deviation from 1:1 segregation is statistically insignificant at the 0.05 level of probability. Ichinose may have the heterozygous constitution of the gene *(11)* since the segregation ratio of lobed and entire leaves in F_1 fits statistically 1:1 in the cross Ichinose X Shiromekeiso (see Table VI). If IRB240-1, No. 3183 and the entire-leaved spontaneous mutant of Kairyonezumigaeshi were induced through a gene mutation, that is, from *L1* to *11*, all F_1 progenies between these mutants and Shiromekeiso should be entire. Contrary to expectations, each of the F_1 progenies produced lobed and entire plants and the segregation ratios were all insignificant, from 1:1 at the 0.05 level of probability.

Consequently, these three mutants are supposed to be not of genic origin according to the data now available. Other mechanisms, e.g. chromosomal aberration, replacement of apical layers or mutable genes, may be responsible for the reversion. In this respect, cytological and histological studies and breeding tests of mutants, including those which show reversion, are now in progress and are to be performed in the near future.

3.2. Induction and selection of die-back-disease-resistant bud mutation by gamma irradiation and inoculation

Mulberry is attacked by many diseases and much of mulberry breeding is concerned with breeding for resistance. In regions of heavy snowfall mulberry is always attacked by die-back disease, *Diaporthe nomurai* Hara. Application of some mercurial preparations is useful for disinfection of the disease; however, these preparations cannot be applied, since environmental pollution by fungicides has recently become a serious problem.

From this point of view, breeding for resistance to die-back disease has, therefore, also become very important. Although there are some resistant varieties to the disease, foliage yield and feeding value in these varieties are generally very low. *Morus bombycis* Koidz, in general, is comparatively more resistant than *M. alba* Lnn. and *M. latifolia* Poilet [14].

It is considered that susceptibility to die-back is likely to be controlled by a single gene (K. Nakajima, personal communication). It is therefore expected that when mulberry is exposed to gamma rays, die-back-resistant bud mutation is induced either by gene mutation or by a small deletion of the chromosome carrying the gene [15].

This experiment has begun to establish a simple and quick method to detect a bud mutation which is highly tolerant or immune against die-back in mulberry, and also to make clear whether any such mutation can be induced by mutagenic treatment with gamma rays.

3.2.1. Materials and methods

The fungus, *Diaporthe nomurai* Hara, was cultured on a sterilized mulberry branch at 23°C in an incubator. Five races isolated from mulberry supplied from the Sericultural Experiment Station, Ojiya, Niigata prefecture were used for the inoculation test. The five races were A-2, B-2, C-2, D-1 and E-1. Two of the races, C-2 and E-1, were isolated from the cultivars Kenmochi and Ichinose grown in the Fukui prefecture. Other races were also isolated from the cultivars Kairyonezumigaeshi, Kenmochi and Nekoyatakasuke. Water suspension of conidia for investigation was prepared according to the method described below.

The pieces of spore horn that covered the surface of the mulberry shoot were removed with platinum wire and dissolved in 0.5 ml of glass distilled water. The conidia suspension was thoroughly stirred with a flash mixer. Four to five mulberry shoots, 20 cm long, were prepared for testing from each mutant strain and the irradiated plant, with careful consideration given to the age and growth of the shoots. These materials were washed in 70% ethanol and the cut ends of these shoots were sealed with wax to keep them from drying (Fig.2).

A drop of conidia suspension was placed on a scratched surface of the shoot with two spots for each race and ten spots per shoot. These inoculated shoots were stored in moistened sawdust and were maintained at 20°C in the incubator. Ten days after inoculation, classification of the resistance was made according to the degree of necrosis at the inoculated point. When necrosis developed at the inoculated point, it was grouped as susceptible and when there was no necrosis, or very little, it was grouped as resistant [16].

Twenty-five induced mutants of the cultivar Ichinose, 18 induced mutants of Kokuso No. 21 and seven induced mutants of Kairyonezumigaeshi were tested. In addition, shoots originating from irradiated buds of two F_1 strains of Hakkokuwase, Tani No. 4693-2 and Tani No. 4711-3 were tested by the same method.

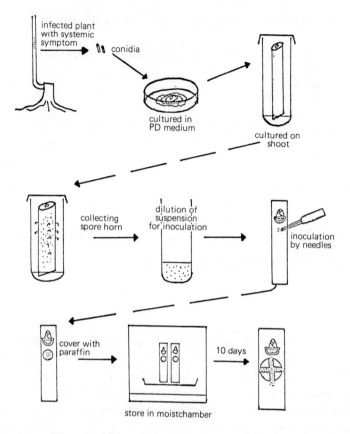

FIG.2. Procedures of the artificial inoculation screening test of the young shoot in die-back disease, Diaporthe nomurai, *of mulberry.*

Irradiation of ^{60}Co gamma rays was applied to detached shoots in the spring of 1972 at 7.5, 10 and 15 kR doses by two different dose rates, 250 and 500 R/h. Irradiated scions were grafted in April and grown in the field for a growing season. Survival of grafts and results of inoculation are shown in Tables VII and VIII.

3.2.2. Results and conclusions

When conidia suspension was applied to mulberry shoots, necrosis was induced only in susceptible mulberry shoots. It is rarely the case that a resistant variety produces a small necrosis. Usually, a resistant variety does not produce the necrosis at the inoculated point. Inoculation is applied

TABLE VII. EXPOSURE AND INOCULATION EXPERIMENT

Total exposure (kR)	Exposure rate (R/h)	No. of investigated plant 4693-2	4711-3	Susceptible	Resistant
15	500	0	0	0	0
15	250	13	9	22	0
10	500	0	3	3	0
10	250	9	11	20	0
7.5	500	0	5	5	0
7.5	250	0	45	43	2[a]
0		57	13	70	0

[a] Not known whether they are truly resistant in the field.

immediately underneath the bud of a shoot, where necrosis is induced more widely in comparison with inoculation over the buds. The inoculation test revealed that none of the 50 mutant strains had any resistance against die-back.

The results of the grafting of irradiated scions are shown in Table VII. The higher the total exposure of gamma rays, the lower the percentage of success of grafting of irradiated scions was observed. At the same total exposure, the higher the exposure rate the lower the percentage of grafting. Two resistant strains among 165 grafts were detected, but it was not ascertained whether they were truly resistant in the field.

So far as the results of these experiments show, it is concluded that inoculation of solution using a vaccination apparatus was most efficient and reliable for selection of resistant mutants.

4.   CHARACTERISTICS OF MUTANTS

4.1. Rooting ability of semi-softwood cuttings in induced mutants

It has already been reported that a number of mulberry mutants have been obtained by gamma-ray irradiation, as well as the rooting ability of hardwood [17]. Mulberry saplings are usually propagated by hardwood cuttings. But semi-softwoods are also employed as the vegetative method of propagation because of ease of cutting and great ability of root initiation, especially in some varieties where it is difficult to initiate the root. Hardwood cutting must be performed up to early summer at the latest and only one cutting from one shoot can be obtained. In the case of semi-softwood, this is not restricted by season with regard to cutting and many cuttings can be obtained from one shoot [18].

From this point of view, the rooting ability of semi-softwood following cutting is an important trait. The rooting ability of two mulberry cultivars, one promising hybrid and 13 induced mutants, is described here.

4.1.1. Materials and methods

Six mutant strains induced from a 5-lobed cultivar of Ichinose, one spontaneous mutant and two induced mutants from the 5-lobed cultivar of Kairyonezumigaeshi, and four mutant strains induced from the promising strain, No. 3052, were used in this experiment. The origin and characteristics of Ichinose, IRB240-1, Kairyonezumigaeshi and the entire-leaved spontaneous mutant of Kairyonezumigaeshi have already been described. The origin and characteristics of the remaining strains are as follows:

IRB240-4:   mutant of IRB240-1, leaves 5-lobed
IRB240-5:   mutant of IRB240-1, leaves 5-lobed
IRB240-6:   mutant of Ichinose, leaves elongated
V 7:   mutant of IRB240-1, leaves entire, sometimes 2-lobed
V 10:   mutant of IRB240-1, leaves 2,3-lobed and entire
215-4:   mutant of Kairyonezumigaeshi, leaves 3-lobed and entire, elongated
511-2:   mutant of Kairyonezumigaeshi, leaves marginally curled, small, wrinkled, dwarf
No. 3052:   hybrid of Ichinose X Kokuso No. 21, leaves entire and lobed, large, thick
IRB240-7:   mutant of No. 3052, short internode
IRB240-8:   mutant of No. 3052, short internode
V 15:   mutant of No. 3052, leaves light green
V 18:   mutant of No. 3052, leaves more lobed.

Thirty cuttings of each were prepared 15–20 cm long, then the leaves were removed, the upper part of two leaves remaining. These materials were put in the propagation bed on 13 June 1975.

Investigation was carried out on the percentage of rooted semi-softwood cuttings, the fresh weight of rooted cuttings, the length of new shoots and the fresh weight of the shoots.

TABLE VIII. SUSCEPTIBILITY RATING IN IRRADIATED CLONES

Total exposure (kR)	Exposure rate (kR/d)	Tani No. 4693-2			Tani No. 4711-3		
		No. of plants	No. of shoots	Susceptibility rating[a]	No. of plants	No. of shoots	Susceptibility rating[a]
21.8	38.0	5	10	1.30	5	12	1.92
17.8	30.0	5	13	1.46	5	11	1.91
14.4	24.5	4	9	1.56	10	9	2.00
12.3	20.5	5	–	–	5	11	1.91
10.4	16.5	5	12	1.50	5	10	2.80
7.7	13.0	3	9	1.56	4	10	2.10
6.7	11.0	4	12	1.50	4	8	2.00
5.9	9.7	5	11	1.18	5	10	1.90
5.2	8.5	4	10	1.40	5	10	2.00
4.7	7.7	4	9	1.33	5	9	2.00
4.2	6.8	4	7	1.57	5	9	1.89
3.7	6.1	5	11	1.09	–	–	–
3.4	5.5	3	8	1.00	5	8	2.00
3.1	5.0	2	7	1.00	5	7	2.00
2.8	4.6	–	–	–	5	8	1.75
7.5	250 R/h	27	57	1.42	34	53	2.04
7.5	500 R/h	30	50	1.50	29	39	1.90

Total exposure (kR)	Exposure rate (kR/d)	Tani No. 4693-2			Tani No. 4711-3		
		No. of plants	No. of shoots	Susceptibility rating[a]	No. of plants	No. of shoots	Susceptibility rating[a]
0	0	10	23	1.43	20	25	1.96
7.5	250 R/h				26	41	2.00
7.5	250 R/h				22	42	1.86

[a] Susceptibility rating = $\dfrac{\text{Number of shoots in each rating}}{\text{Total number of shoots}}$

TABLE IX. PERCENTAGE OF ROOTED SEMI-SOFTWOOD CUTTINGS, LENGTH OF SHOOT AND WEIGHT OF ROOT AND SHOOT IN THE CUTTINGS OF TWO MULBERRY CULTIVARS, 1 HYBRID AND 13 GAMMA-RAY INDUCED MUTANTS DERIVED FROM THEM

Variety and mutant[a]	Percentage of rooted semi-soft-wood cuttings (%)	Fresh weight of root per cutting (g)	Length of new shoot (cm)	Fresh weight of shoot per cutting (g)
Ichinose (original)	53	1.17	9.12	3.45
IRB240-1	64	1.08	9.20	5.32
IRB240-4	60	1.73	8.64	7.96
IRB240-5	64	2.22	4.50	2.24
IRB240-6	36	1.58	13.01	6.08
V 7	40	1.72	15.75	11.47
V10	80	2.04	12.90	9.10
Kairyonezumigaeshi (original)	96	2.70	8.50	4.08
Entire-leaved spontaneous mutant of Kairyonezumigaeshi	68	2.58	4.10	4.12
215-4	16	2.85	8.00	4.11
511-2	24	1.90	5.50	1.30
No. 3052 (original)	32	4.80	7.00	4.50
IRB 240-7	42	2.00	11.00	5.32
IRB 240-8	28	2.32	10.30	2.24
V 15	32	1.30	5.13	5.75
V 18	44	2.48	14.25	8.43

[a] IRB240-1, IRB240-4, IRB240-5, IRB240-6, V 7 and V 10 were derived from Ichinose; 215-4 and 511-2 were derived from Kairyonezumigaeshi, IRB240-7, IRB240-8, V 15 and V 18 were derived from No. 3052.

4.1.2. Results and conclusions

The percentage of rooted semi-softwood cuttings of the originals and 13 mutant strains studied is given in Table IX, as well as the fresh weight of root and shoot and the length of semi-softwood. As for the rooting ability, Ichinose and its mutant strains had a relatively low percentage of 50%, whereas in V 10 the percentage was 80%. The spontaneous mutants of Kairyonezumigaeshi, 215-4 and 511-2, had a low percentage of 16−32%, whereas in Kairyonezumigaeshi the percentage was 96%. No. 3052 and the mutants of No. 3052 had a comparatively low percentage and there was no difference between them. The mutants derived from Ichinose and No. 3052, which were obtained by gamma irradiation and re-irradiation, did not exhibit any decrease in rooting ability, but the mutants derived from Kairyonezumigaeshi decreased in rooting ability when compared with the original. Growing vigorousness was generally more decreased than the original when the plants were treated by gamma irradiation. There were some mutant strains that did not show any disability of rooting initiation of shoots, namely IRB240-1, IRB240-5, V 10 and IRB240-7.

As a result of this experiment, it is considered possible that useful mutants which show high rooting ability can be selected by gamma irradiation.

4.2. Electron-microscopic study of the trichome and idioblast

The shape of the idioblast, which is distributed on the leaf, and the length and density of the trichome are specific to the mulberry species; they are the most reliable features for classification of mulberry trees [19, 20]. Previously we investigated several mulberry mutants for some traits, such as the rooting ability of cuttings [21]. The trichome and idioblast of mulberry mutants are dealt with here.

4.2.1. Materials and methods

Six mulberry varieties and 24 mulberry mutant strains derived from these varieties by gamma irradiation were used. Winter buds, leaves and the epidermis of shoots were collected in the early spring of 1979. These materials were dried with freeze-drying methods and/or with a critical point-drying method. For microscopic observation the third or fourth scale from the surface of the winter bud was taken off and the leaves and epidermis of shoots were cut off in small pieces. Most of these materials were dried with the freeze-drying method, but some were dehydrated by ethanol and acetone, following critical point-drying with a Hitachi HPC-2. The scale bracts, pieces of leaf and the epidermis of shoots were mounted on an aluminium bed with adhesive tape and were then coated with gold in a vacuum and viewed in a Hitachi model S-310A scanning electron microscope at an accelerating voltage of 5 kV.

(text cont. on p.278)

TABLE X. CHARACTERISTICS, QUANTITATIVE MORPHOLOGICAL CHARACTERS OF IDIOBLAST AND PUBESCENT OF SCALE

Mutant strain and variety	Original variety	Characteristics	Diameter of idioblast		Length of stomata (μm)	Length of trichome	
			Upper epidermis (μm)	Lower epidermis (μm)		Scale (μm)	Shoot (μm)
Kairyonezumigaeshi		Five-lobed	31	60	12	50	50 thin
Ichinose		Five-lobed	34	41	14	50	90
No. 3052		Entire	20	40	12	47	80
Keguwa		Entire	14	8	6.5	92	200
Mon noi		Entire	41	38	8	35	39
M. nigra		Entire, lobed	105	69	28	50	91
215-4	Kairyonezumigaeshi	Elongated	68	30	15	67	68
469	Kairyôhezumigaeshi	Entire, lobed	44	39	16	50	no trichome
472-5	Kairyonezumigaeshi	Elongated	30	36	18	35	65
478	Kairyonezumigaeshi	Small leaf	39	35	16	58	120
484	Kairyonezumigaeshi	Semi-dwarf	25	24	13	36 thin	38 thin
Maruba Kairyonezumigaeshi	Kairyonezumigaeshi	Entire	51	34	16	70	120 long
IRB 240-1	Ichinose	Entire	23	75	15	55 thin	70
115	Ichinose	Entire, many lobed	35	34	20	26 bifurcate	60 thin
146	Ichinose	Entire, thick	—	53	13	42 thin	28
156	Ichinose	Semi-dwarf	39	20	13	44 thin	70 long, dense
170	Ichinose	Elongated	45	48	18	70	50
434-1	Ichinose	Elongated	34	40	12	49 dense	17 short, thin
442	Ichinose	Short internode	—	39	11	47	53 dense

Mutant strain and variety	Original variety	Characteristics	Diameter of idioblast		Length of stomata (μm)	Length of trichome	
			Upper epidermis (μm)	Lower epidermis (μm)		Scale (μm)	Shoot (μm)
V 2	IRB 240-1	Elongated	–	–	–	32	120
V 4	IRB 240-1	One or two-lobed	44	32	16	38 thin	60
V 7	IRB 240-1	One-lobed	25	25	14	37 thin	60
V 9	IRB 240-1	Elongated	36	39	15	38	37 dense
V 10	IRB 240-1	One-or two-lobed	39	37	12	41	35 short
V 13	No. 3052	Short internode	44	39	12	43 dense	80
V 14	No. 3052	Light green	30	40	16	40 thin	135
V 15	No. 3052	Light green	51	no idioblast	21	30 slender	45 thin
V 17	No. 3052	five-lobed	36	39	13	35	55 thin
V 18	No. 3052	five-lobed	32	39	13	34 dense	43

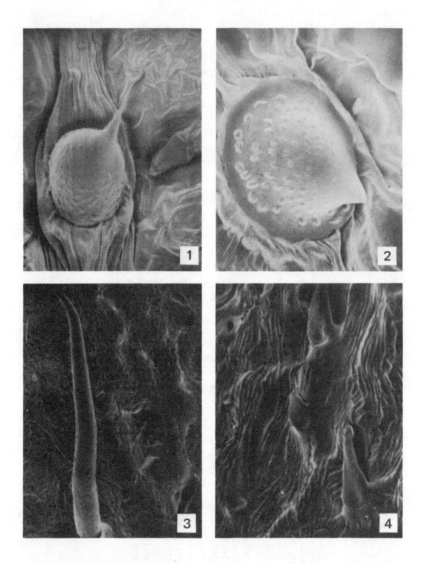

FIG.3. Kairyonezumigaeshi showing: 1. Idioblast of upper epidermis on leaf and stomata;
2. idioblast of lower epidermis; 3. trichome on shoot; 4. trichome on scale.

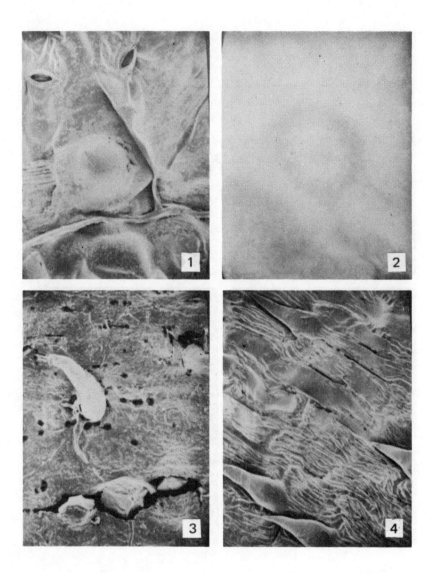

FIG.4. 484 showing: 1. Flat surface of idioblast; 2. idioblast of lower epidermis; 3. trichome on shoot; 4. trichome on scale.

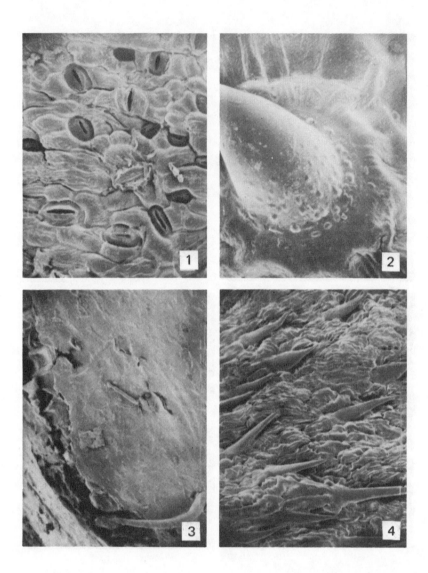

FIG. 5. V 15 showing: 1. No idioblast on leaf; 2. idioblast of lower epidermis; 3. trichome on shoot; 4. trichome on scale.

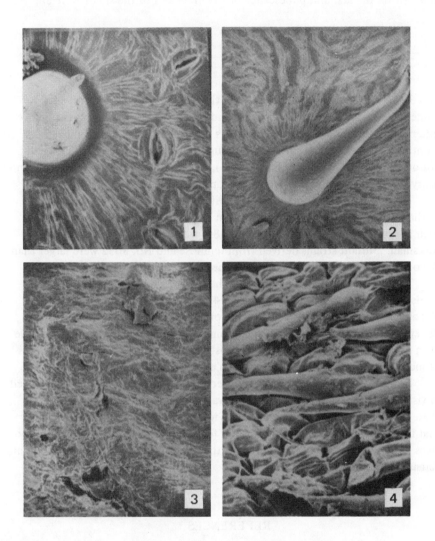

FIG.6.   469 showing: 1. Idioblast of upper epidermis on leaf and stomata; 2. idioblast of lower epidermis; 3. no trichome on shoot; 4. trichome on scale.

Micrographs were made with a Polaroid Type 52 positive film at magnifications from 750× to 3000×. For study and measurement, magnifications of 1500× idioblast on the leaf and pubescence of the bract and of the shoot were applied.

4.2.2. Results and discussion

The results of microscopic observations are shown in Table X and Figs 3–6. The shape of the idioblast on the leaf surface has multiformity. Some of the shapes have a hemisphere and others have a process on the hemispherical idioblast. Most of the idioblasts have many small dots similar to an eruption on the spherical surface. Although Kairyonezumigaeshi has a small dot on the surface of the idioblast (Fig.3), 215-4 and 484, which derived from Kairyonezumigaeshi, have a flat surface on the idioblast (Fig.4).

There are a small number of idioblasts in 484 and 469. The diameter of the idioblast on the upper epidermis in mutant strains was large compared with the original, except for 472-5 and 484, whereas the abaxial diameter was small. Mutant strains derived from Ichinose and No. 3052 were larger than the originals. V 15 had no idioblast in the abaxial epidermis (Fig.5).

The trichome on the epidermis of the shoot and scale are also shown in Table X. The length of the trichome on 478, Maruba Kairyonezumigaeshi, V 2 and V 14 was much longer than the originals; 469 had no trichome on the epidermis of the shoot (Fig.6). A number of mulberry mutant strains have been obtained by gamma irradiation, most of them being selected on the basis of morphological or visible changes, such as leaf shape, leaf size.

From the result of microscopic observation, these mutants not only varied in visible changes from the original, but varied in invisible changes, namely, these mutant strains have polymorphous idioblasts. Katsumata confirmed that the shape of the idioblast is a useful feature for classification of mulberry trees [19]. Further study is required on the relationships between this result and Katsumata's results.

REFERENCES

[1] HAZAMA, K., Breeding mulberry trees, Jpn. Agric. Res. Quart. 3 (1967) 15.
[2] HAZAMA, K., Varietal differences of mutation rate and mutation spectrum after acute gamma irradiation in mulberry, J. Sericult. Sci. Japan 39 (1968) 194.
[3] FUJITA, H., NAKAJIMA, K., Re-treatment of induced mulberry mutants with gamma rays, Gamma Field Symp. 12 (1973) 49.
[4] NISHIDA, T., Studies on the radiosensitivity of fruit trees. I. Differences of radiosensitivity within the fruit trees under chronic gamma ray irradiation, Paper presented at the Spring Convention of Japan Soc. Hort. Sci. (1965).

[5] NAKAJIMA, K., Induction of sports in roses by gamma ray irradiation, Gamma Field Symp. 4 (1965) 55.

[6] HAZAMA, K., KATAGIRI, K., TAKATO, S., Varietal difference of radiosensitivity and bud mutation of mulberry tree in a gamma irradiation field, J. Sericult. Sci. Japan. 37 5 (1968) 427.

[7] FUJITA, H., TAKATO, S., An entire leaf mutant in mulberry, Inst. Radiat. Breed. Tech. News No. 5 (1970).

[8] FUJITA, H., Bud mutation of mulberry with chronic irradiation in the gamma field, Gamma Field Annual Report (1975) 21.

[9] GAUL, H., "Use of induced mutants in seed propagated species", Mutation and Plant Breeding, National Academy of Sciences National Council (1961) 206.

[10] DOMMERGUES, P., «Action des rayons gamma sur les bourgeons de la variété de poirier Max Red Bartlett», Effects of Ionizing Radiation Seeds (1961) 581.

[11] MEHLQUIST, G.A.L., SAGAWA, Y., "The effect of gamma radiation on carnations", Proc. 16th Int. Hort. Congr. 4 (1962) 10.

[12] BOWEN, P., CAWSE, P.A., DICK, M.J., The induction of sports in chrysanthemums by gamma radiation, Radiat. Bot. 1 (1962) 297.

[13] AIDA, E., NAKAJIMA, K., MATSUSHIMA, M., A cross experiment regarding the breeding behavior of leaf shape in mulberry, J. Sericult. Sci. Japan 31 (1962) 182.

[14] AOKI, K., Aetiological studies of mulberry-blight, Dogare-disease, Bull. Sericult. Exp. Sta. XII (1944) 245.

[15] NYBOM, N., KOCH, A., "Induced mutations and breeding method in vegetatively propagated plants", The Use of Induced Mutation in Plant Breeding; Report of FAO/IAEA Technical Meeting Rome, 1964, Pergamon Press, Oxford (1965) 661.

[16] MIYAYAMA, K., OKABE, T., Studies on the test for the resistance of die-back disease by inoculation of conidia, J. Sericult. Sci. Japan 41 (1971) 230.

[17] KUKIMURA, H., et al., "Genetical and physiological studies on the induced mutants with special regard to effective methods for obtaining useful mutants in perennial woody plants", Improvement of Vegetatively Propagated Plants Through Induced Mutations, Tech. Doc. IAEA-173 (1975) 83 (unpublished).

[18] HONDA, T., Studies on the propagation of mulberry trees by cutting, Bull. Sericult. Exp. Sta. 24 (1970) 133.

[19] KATSUMATA, F., Relationship between the length of styles and the shape of idioblasts in mulberry trees, J. Sericult. Sci. Japan, 41 5 (1971) 387.

[20] HOTTA, T., Kuwa hen, Yoken-do Co. Ltd. (1951) 24.

[21] KUKIMURA, H., et al., Genetical, cytological and physiological studies on the induced mutants with special regard to effective methods for obtaining useful mutants in perennial woody plants, Progress Report, IAEA, 1975 (unpublished).

STUDIES ON CYTOLOGICAL, PHYSIOLOGICAL, AND GENETIC CHARACTERISTICS IN SOMATIC MUTANT STRAINS OF SUGI (*Cryptomeria japonica* D. Don)*

T. MAETA
Kanto Forest Tree Breeding Institute, MAFF,
Mito

M. SOMEGOU, K. NAKAHIRA
Forestry and Forest Products Research
    Institute, MAFF,
Tsukuba

Y. MIYAZAKI
Kyushu University Forests,
Fukuoka

T. KONDO
Institute of Radiation Breeding, NIAS, MAFF,
Ohmiya, Ibaraki-ken,
Japan

Abstract

STUDIES ON CYTOLOGICAL, PHYSIOLOGICAL AND GENETIC CHARACTERISTICS IN SOMATIC MUTANT STRAINS OF SUGI *(Cryptomeria japonica* D. Don).

From microscopic observation of the pollen of induced mutant strains in Sugi *(Cryptomeria japonica* D. Don), it was found that there were large differences in pollen fertility among the mutant strains, and that it deviated year to year from the mother plants. The large differences in frequency of sterile pollen among mutant strains depended on the genetic characteristics of each mutant strain. Higher frequencies of sterile pollen were observed at the terminal part of branchlets in some mutant strains, and this was considered to be induced by the lateness of flower-bud formation at low temperature conditions in late summer. Delayed formation and gibberellic acid treatment applied for flower induction resulted in low fertility and abnormality of pollen in mutant strains. Chromosome aberration in mutant strains was caused either by gamma irradiation or by some mutational events that responded to environmental conditions. In the former case, aberration might have been maintained for a long period through vegetative propagation. Some of the irregularities were due to mitotic cell division, because cells with micronuclei at the pacytene stage in pollen mother cells and with fragments at MI were observed. Somatic mutability of Kuma-sugi mutants after re-irradiation was investigated. From waxless mutants morphological somatic mutations, which have fat or stout stems and thick and short needles, were frequently produced, whereas from morphological mutants the lowest somatic mutation frequency was induced. In some mutant strains higher rooting ability than the mother plants was found, and the possibility of character improvement was pointed out.

* Research supported by the IAEA under Research Contract No. 1336.

1.   INTRODUCTION

Sugi, *Cryptomeria japonica* D. Don, is one of the most important forest tree
species. It has wide natural distribution in Japan and has been planted for wood
production. To enlarge the range of variation, induction of somatic mutation in
Sugi has been carried out by Ohba [1, 2] and Ohba and Murai [3, 4]. About
90 somatic mutant strains derived from important forest trees, such as *C. japonica*,
Pinus densiflora, *Chamaecyparis obtusa* and *C. pisifera*, were planted at the Institute
of Radiation Breeding. The characteristics of some mutant strains were nearly the
same as the segregated seedlings from self- or cross-pollination of the mother plants.
Mutants which have chlorophyll anomalies are certainly considered to derive from
gene mutation, as Aliston and Sparrow [5], Nybom and Koh [6] and others have
pointed out that deletion of dominant genes resulted in frequent somatic mutation.
However, there are morphological mutants that not only have a single mutated
character, but also have several characters that are drastically mutated. These
morphological mutant strains are inferior in height and diameter growth and
show a wide variation in cross-fertility [7]. Somatic mutants in Sugi are usually
classified into chlorophyll, morphological and waxless mutants, but their
characteristics, except visible characters and seed fertility, have not been clarified.
The re-irradiation effect on Sugi mutant strains, rooting ability, pollen fertility
and abnormal division in meiosis are reported on here, with regard to the Sugi
reproductive methods. Observations of peroxidase isozyme phenotypes were
carried out, but the mother plant, used as the control, of each mutant strain was
not suitable because of clone complex. Thus, studies on the variation of isozyme
phenotypes have been omitted here.

2.   CYTOLOGICAL STUDIES ON INDUCED MUTANTS OF SUGI

2.1. Abnormal division in meiosis and pollen fertility in some Sugi mutant strains

In Sugi drastic morphological mutations, such as dwarf, abnormal needle
shape and waxless needles, were induced by acute or chronic gamma-ray
irradiation (Table 1). Except for a few strains most of these mutant strains
showed low pollen fertility. In our experiment male flowers were collected from
material grown in the greenhouse and the field, and meiotic division and pollen
fertility were observed in several mutant strains.

2.1.1. Materials and methods

Observations on pollen mother cell division of Sugi were conducted on seven
mutant strains (IRB601-1, 601-3, 601-6, 601-14, 601-22, 601-23 and 601-32)

and their controls. Induction of flower-bud formation was performed by leaf-surface spraying of a 100 ppm gibberellic acid solution. Male flowers were removed, fixed in Carnoy's solution, stained with acetocarmine and squashed. Pollen mother cell division was observed in pre-matured male flowers in September 1973. Mature pollen was observed by staining with acetocarmine in February 1974.

2.1.2. Results and discussion

Meiotic cell division of mutant strains grown in the greenhouse and the field was normal and there were no differences between the original varieties and the mutants. A few abnormal meiotic cell divisions were observed in the pilot test, but in our observations of pollen mother cells, their abnormal division, such as lagging chromosomes, multipolar division and micronuclei, was not observed. At the MI stage all the mutant strains examined formed 11 pairs of normal, bivalent chromosomes and their pairing was complete. At the MII stage no abnormal divisions, such as chromosomal bridges and fragments, were observed. In the mutant strains IRB601-3, 601-6, 601-22 and 601-14 some of the pollen mother cells ceased dividing during meiotic cell division and large quantities of abortive pollen were produced. On the other hand, in the control plants IRB601-23 and IRB601-32 little abortive pollen was observed and meiotic division was considered to be normal (Table 2). Abnormal pollen grains produced in the mutant strains IRB601-1, 601-3, 601-6, 601-14 and 601-22 were smaller in size and had thicker pollen grain walls than in normal pollen grains; abnormal pollen was presumably caused by the development process after the pollen tetrad stage.

2.2. Abnormal pollen formation in Sugi mutant strains under different environmental conditions

2.2.1. Materials and methods

Observations on pollen fertility and chromosomes were carried out on nine mutant strains collected from many mutant strains. The mutant strains IRB601-1, 601-3, 601-13, 601-22, 601-23, 601-32, 601-65 and 601-70 were used and their characteristics are shown in Table 1. These mutant strains were propagated vegetatively and were planted in the mutant bank and in pots in the greenhouse. Girdling and/or gibberellic acid treatment (100 ppm aqueous solution) were applied to the materials to induce flower buds. A comparison was made of the pollen fertility of materials in the glasshouse and in the field. Male flowers of the mutant strains were collected from the terminal and basal parts of the branchlets. By girdling, however, no male flower bud was induced in any of the mutant strains, hence it could not be established whether abnormal

Table 1. Characteristics, exposed doses, exposed year and detection of Sugi mutant strains

IRB No.	Gamma ray irradiation				Characteristics
	Methods	Exposed dose (R)	Exposed in	Detected in	
601- 1	Acute	1440	1962	1963	Slender branches, adnate needles
601- 2	"	"	"	"	Slender branches, adnate, twining needles
601- 3	"	"	"	"	Twining needles in sprouting
601- 5	Chronic	540	1962	1963	Long needles
601- 6	Acute	600	1963	1965	Dwarf
601- 9	"	580	1965	1966	Dwarf, twining needles
601-11	"	"	"	"	Dwarf, twining needles
601-13	"	600	1963	1964	Twining needles
601-14	"	"	"	"	Argute needles
601-15	"	700	1964	1965	Straight, fine, adnate needles
601-16	Chronic	-	1962	1966	Long, straight needles
601-17	Acute	580	1965	1966	Argute needles
601-17m	"	"	"	"	Slender branches
601-18	"	800	1964	1965	Dwarf. short, argute needles
601-21	Chronic	1456	1962	1963	Waxless
601-22	"	"	"	"	Dwarf. waxless
601-23	Acute	2880	1962	1964	Waxless
601-24	"	600	1963	1964	Waxless
601-26	"	1200	1964	1965	Waxless
601-32	"	580	1965	1966	Waxless
601-33	"	"	"	"	Waxless
601-34	"	"	"	"	Waxless, light green needles
601-35	"	"	"	"	Waxless
601-65	Chronic	4120	1962	1967	Dwarf. fine short needles
601-80	-	-	-	-	Dwarf. argute, shoet needles
WPL147	-	-	-	-	White leaves in sprouting

Table 2.   Pollen fertility of mutant strains of Sugi (1973)

IRB No.	Number of pollen observed	Number of fertile pollen (%)	Number of sterile pollen (%)
601- 1	206	122(59.2)	79(38.4)
601- 3	205	38(18.5)	119(58.0)
601- 6	206	22(10.7)	139(67.5)
601-14	150	28(17.7)	116(77.3)
601-22	340	124(36.5)	197(57.9)
601-23	172	158(91.9)	14(8.1)
601-32	169	159(94.1)	10(5.9)
Control	303	300(99.0)	3(0.1)

pollen was a direct result of the gibberellic acid treatment or not. The chromosome and pollen observations were carried out by the same methods described in sub-section 2.1.1.

2.2.2. Results and discussion

At the MI stage 11 normal, firmly pairing bivalents were observed in all mutant strains. At AI and TI chromosomes moved to each pole normally and through MII, AII and TII, i.e. they formed normal pollen tetrads. No differences in the frequency of sterile pollen in the mutant strains were recognized between the field and the glasshouse. Mutant strains that had a high frequency of sterile pollen in the glasshouse also showed a high frequency when grown in the field. In IRB601-32, 601-65 and 601-70 the differences in frequency of sterile pollen formed at the basal and terminal parts of the branchlets were observed both in the field and in the glasshouse. In comparison with the 1973 data yearly differences of the frequency were observed in IRB601-3, but in some mutant strains there were no yearly differences, e.g. IRB601-23. From these results it was thought that the large differences in frequency of sterile pollen among the mutant strains depended on the genetical characteristics of each mutant strain. Higher frequencies of sterile pollen were observed at the terminal parts of the branchlets in some mutant strains and it was considered that they were induced by the lateness of flower-bud formation at low temperature conditions in late summer (Table 3).

2.3. Abnormal pollen formation by a high concentration treatment of gibberellic acid solution

The aim of this experiment was to ascertain whether mutation itself or the effects of the gibberellic acid solution treatment was responsible for low pollen fertility.

Table 3. Frequency of sterile pollen (1974)

IRB No.	In field		In glass house	
	Basal part of branchlet (%)	Terminal part of branchlet (%)	Basal part of branchlet (%)	Terminal part of branchlet (%)
601- 1	11.0	17.6	18.8	14.9
601- 3	57.9	55.8	48.0	49.5
601- 6	69.6	68.8	-	-
601-13	2.7	2.1	-	-
601-14	76.8	73.1	-	-
601-22	59.5	66.1	61.1	61.3
601-23	3.0	4.4	0.8	1.1
601-32	69.5	80.1	-	-
601-70	44.4	92.3	-	-
601-65	-	-	6.8	10.8

2.3.1. Materials and methods

The mutant strains used in this microscopic observation were IRB601-1, 601-3, 601-14 and 601-22, with the original plant (Kuma-sugi) as the control. These mutant strains and the control plant were treated with 10, 30, 50, 100 and 200 ppm of gibberellic acid solution to induce flower-bud formation. The branchlets of each material were dipped into each gibberellic acid solution for approximately 1 min. This treatment was carried out twice between late June and early July 1976. After the treatments male flower buds formed in the first 10 days of August, started meiotic division in the middle of September and formed mature pollen by the middle of October. On 10 November the male flowers of these mutant strains were collected for microscopic observation of the pollen. It was thought to have matured by this time, even in mutant strains that had late maturation. Male flowers were collected from both the terminal and basal parts of the branchlets to examine the differences in earliness of flower-bud formation and pollen fertility.

2.3.2. Results and discussion

Three mutant strains, IRB601-3, 601-14 and 601-22, formed male flowers only in the terminal parts of the branchlets, not in the basal parts. In the control plant, the frequency of fertile pollen was very high and the percentages were more than 97% in all plants treated with gibberellic acid (Table 4), whereas pollen fertility in the mutant strains had large differences among the strains. Furthermore,

Table 4. Effect of gibberellic acid concentration on pollen fertility of Sugi mutant strains

IRB No.	GA treat-ment(ppm)	Collected part	Pollen observed	Fertile pollen(%)	Micro (%)	Empty (%)	Gigantic (%)	Other abnormal pollen (%)
601- 1	10	1*	798	78.5	13.7	5.6	0.5	1.6
		2*	606	51.7	45.7	2.5	0.	0.2
	30	1	1702	60.5	26.6	12.0	0.7	0.3
		2	1382	61.2	33.4	3.4	0.4	0.4
	50	1	1975	77.9	17.3	4.7	0	0.1
		2	1131	72.5	20.9	5.8	0.1	0.7
	100	1	1841	72.4	22.1	5.2	0.1	0.2
		2	1668	25.2	72.7	1.0	0.1	0.7
	200	1	-	-	-	-	-	-
		2	1300	0.9	98.8	0.	0	0.2
601- 3	10	2	1790	58.4	21.8	19.7	0.	0.1
	30	2	3082	56.9	24.2	18.7	0.2	0.03
	50	2	1765	49.4	39.3	11.2	0.1	0
	100	2	1638	7.6	91.7	0.1	0.5	0
	200	2	1782	56.0	37.6	6.0	0.3	0.1
601-14	10	-	-	-	-	-	-	-
	30	-	-	-	-	-	-	-
	50	2	1189	43.7	46.4	8.5	0.2	1.2
	100	-	-	-	-	-	-	-
	200	2	718	0.	84.8	15.1	0	0
601-22	10	2	1247	42.3	38.5	16.8	2.1	0.3
	30	2	1703	36.9	53.0	8.6	1.1	0.4
	50	2	1737	9.0	75.2	15.3	0.4	0
	100	2	1518	52.2	23.3	17.7	6.9	0
	200	2	1371	3.9	87.5	7.5	0.2	0.9
Control	10	1	739	98.4	0.8	0.8	0	0.
		2	1020	98.2	0.8	1.0	0	0
	30	1	1262	98.8	0.8	0.4	0	0
		2	1322	98.0	1.2	0.8	0	0
	50	1	1517	97.9	1.1	1.1	0	0
		2	1215	98.1	1.0	0.9	0	0
	100	1	1408	98.2	1.0	0.8	0	0
		2	1605	97.8	1.8	1.4	0	0
	200	1	1590	97.8	0.8	0.5	0	0
		2	1018	98.5	0.1	0.5	0	0

* 1; Basal part of brachlet with male flowers.
2; Terminal part of branchlet with male flowers.

reduction of fertility was larger because of the higher concentration of gibberellic acid. No differences of pollen fertility in either part of the branchlets were detected in the control, but in IRB601-1 pollen fertility was reduced greatly by the high concentration of gibberellic acid. It may be concluded that flower-bud formation and development in these mutant strains took longer than the control because they had smaller and immature pollen grains. In IRB601-22, a higher frequency of gigantic pollen grains was observed than in the other mutant strains; this might be caused by an obstruction of division at MII or cellularization at the pollen tetrad stage.

Kuma-sugi had a higher response to the gibberellic acid treatment and easily formed many male and female flowers, whereas the mutant strains had a low sensitivity to gibberellic acid, only a few flower buds were formed and their development was incomplete. It was presumed that the flower and pollen formations of these mutant strains were easily influenced by environmental conditions, since a lower temperature was recorded at the meiotic stage of Sugi in 1976. As a result, pollen fertility in the terminal parts of the branchlets was reduced by delayed development of the pollen mother cells. As was observed in the control, there may be no direct effects of gibberellic acid to abnormal pollen formation. It was, however, considered that delayed formation and development of flowers was caused by the remaining effects of a high concentration of gibberellic acid, resulting in the low fertility and anomaly of pollen in mutant strains.

2.4. Abnormal movement of chromosomes in Sugi mutant strains

Cytological studies on pollen fertility and the behaviour of chromosomes at meiosis in Sugi mutant strains induced by gamma irradiation have been reported since 1973. Pollen fertility by microscopic observation was very low, although anomaly at meiosis was rarely observed. Therefore, the relation between irregularity in pollen formation and the concentration of the gibberellic acid solution applied to induce flower buds was examined in 1976. It was concluded that the delayed formation of male flowers at low temperatures caused abnormal pollen. In the control plants (Kuma-sugi), however, low cross- or self-fertility is generally known, therefore meiosis in mutant strains orginating from Kuma-sugi (local variety of Sugi, 2n=22) was examined, as well as chromosome aberration and its behaviour.

2.4.1. Materials and methods

The 13 mutant strains examined originated from Kuma-sugi and their characteristics are noted in Table 1. The original plant and the mutant strains IRB601-1, 601-2, 601-3, 601-6, 601-9, 601-17, 601-17m, 601-21, 601-22,

FIG.1. Normal 11 bivalent chromosome at MI (IRB601-24).

601-23, 601-24, 601-33 and WPL 147 were treated with a 100 ppm gibberellic acid solution to induce flower buds. Procedures for observation were the same as in sub-section 2.1.1. Observations were carried out on the cells at the late prophase of MI, AI, MII, AII and at the tetrad stage.

2.4.2. Results and discussion

Eleven normal bivalent chromosomes (Fig.1) and abnormal pairing of chromosomes at the late prophase of MI were observed; the anomalies were as follows (Fig.2):

(i) Bivalent delayed normal chromosome movement situated outside the nuclear plate (1_{II}, Fig.2(a))
(ii) Univalent (Fig.2(b))
(iii) Quadrivalent (Fig.2(c))
(iv) Chromosome fragment (Fig.2(d)).

The partially homologous chromosomes were located mainly in other normal chromosomes; however, some of them were often situated outside the nuclear plate, but they were still classified as partially homologous. Lagging chromosomes and chromosome bridges at AI and abnormal chromosomes at the MII and AII stage, as already mentioned, were observed (Fig.3). Normal chromosome distributions were represented as 11:11 (Fig.4) at MII and 11:11:11:11 at AII. Lagging chromosomes situated outside the nuclear plate at MI were represented as 10:10:2 and laggards on the nuclear plate at MII were represented as 10:1:11.

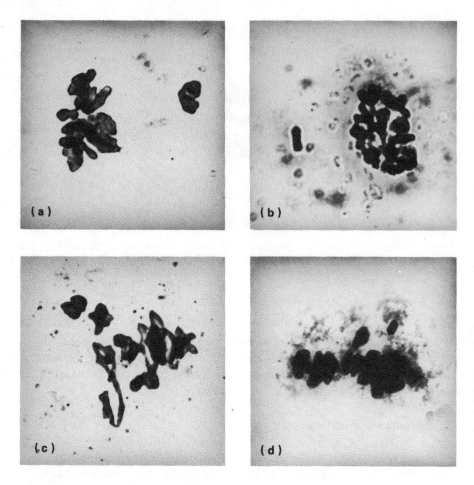

FIG.2. Meiotic abnormalities at MI: (a) Bivalent situated outside nuclear plate (original plant); (b) Univalent (IRB601-9); (c) Quadrivalent (IRB601-2); (d) Chromosome fragment at MI (IRB601-2).

At the pollen tetrad stage cells were classified into either normal tetrad or abnormal cells with one or two micronuclei. Results of observations at each stage in the original and some of the 13 mutant strains observed are shown in Table 5.

In the original plant the frequency of normal cells with 11_{II} and other abnormal cell frequencies, such as cells with partially homologous chromosomes and the bivalent situated outside the nuclear plate at the late prophase of MI, were 88.71 and 11.28%, respectively. The frequency of cells with laggards at AI was 11.53%. Unequal chromosome distribution and laggards were also observed at MII and AII. Cells with micronuclei at the pollen tetrad stage were 6.25%.

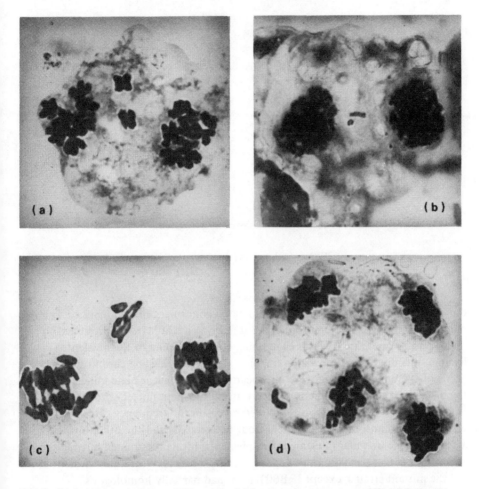

FIG.3. Meiotic abnormalities at MII to AII: (a) Lagging chromosomes at MII (IRB601-17); (b) Chromosome fragment between daughter nuclei (IRB601-33); (c) Lagging chromosomes at early AII (IRB601-17); (d) Lagging chromosomes at AII (IRB601-17).

Thus, some anomalies in chromosome behaviour were found even in the original plant, which was assumed to be normal.

In the mutant strains partially homologous chromosomes, bivalents situated outside the nuclear plate, univalents, quadrivalents and chromosome fragments at the late prophase of MI, and other anomalies, such as unequal chromosome distribution and laggards at MII, were observed (Fig.5). Considerable frequencies of laggards, fragments and chromosome bridges at AI and AII (Fig.6), as well as micronuclei at the tetrad stage were observed (Fig.7). The small number of cells

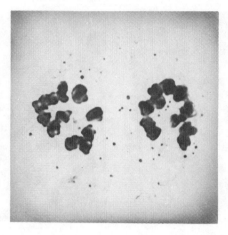

FIG.4.   Normal MII (original plant).

observed at each stage gave rise to the unequal abnormal cell frequencies. The
partially homologous chromosomes and quadrivalents at MI were presumed to
divide and to orientate towards each pole. However, some of the bivalents
located outside the nuclear plate remained and formed lagging chromosomes.
Some of the univalents that were not located on the nuclear plate joined one of
the daughter nuclei and some remained outside the main nuclei. From these
results it can be seen that there were differences in the frequency of cells with
abnormal chromosomes and the behaviour between MI and the other bivalents.
Because of the irregular movement of univalents, bivalents situated outside the
nuclear plate had partially homologous chromosomes and were quadrivalent at MI.
All the mutant strains, except IRB601-17m, had partially homologous
chromosomes and were quadrivalent. Chromosome fragments were observed in
four mutant strains (IRB601-3, 601-9, 601-22 and 601-33) at MI and unequal
chromosome distribution was found in all the mutant strains at MII and AII.
The total frequencies of abnormal cells at each stage are shown in Table 6.
Unequal frequencies at AI among mutant strains were due to the insufficient
number of cells under observation.

Among the mutant strains examined two (IRB601-21 and 601-22) were
induced by chronic gamma irradiation and the rest by acute irradiation. There
was no relation between the exposed dose or the methods of exposure and the
abnormal cell frequencies. It has already been mentioned that pollen fertility in
the mutant strains IRB601-1, 601-6 and 601-22 was very low and that the micro-
pollen frequency was the largest component of the sterile pollen. In our
observation, however, the highest frequency of cells with micronuclei at the
tetrad stage was 33.33% (IRB601-17m). It was expected that there would be

some yearly differences of micro-pollen formation in the mutant strains. The original strain, Kuma-sugi, is one of the strains that has a rather high mutation frequency among several local varieties planted in the gamma-field and also has a low fertility of self-pollinated seeds. From these characteristics it was expected to have some anomalies. Our observations clarified the cause of low fertility of self-pollinated seeds of Kuma-sugi. Considerable frequency of normal cells at the MI stage in mutant strains was observed and we could not characterize them by their chromosomal aberration because of the low frequencies of partially homologous chromosomes, unsyncronized chromosomes, univalents, quadrivalents or fragments. From our observation, it is considered that chromosome aberrations in mutant strains are caused either by gamma irradiation or by some mutational events which respond to environmental conditions. In the former case, aberration might have been maintained for long periods through vegetative propagation. It is presumed that some of the irregularities were due to mitotic cell division, because cells with micronuclei at the pacytene stage in pollen mother cells (Fig.8) and fragments at MI were occasionally observed.

3.   INDUCTION OF SOMATIC MUTATION IN MUTANT STRAINS OF KUMA-SUGI (LOCAL VARIETY *Cryptomeria japonica* D. Don) BY GAMMA RE-IRRADIATION

It was clarified that Kuma-sugi (local variety of Sugi) showed a high frequency of induced mutation by gamma irradiation and differences of mutation frequency of the pruning treatment after irradiation [8]. Furthermore, the present experiment was performed to obtain a much higher mutation frequency and to examine the kinds of somatic mutation induced by gamma re-irradiation.

3.1. Materials and methods

The materials used in this experiment were eight mutant strains (IRB601-1, 601-3, 601-6, 601-15, 601-17, 601-21, 601-23 and 601-26), together with clones without any morphological changes after mutagenic treatment. Their characteristics, doses and years when exposed and detected are shown in Table 1. Mutants and their control plants (without any morphological changes after mutagenic treatment) were raised by cuttings from each Kuma-sugi which produced somatic mutations by gamma irradiation; they were grown in pots with a 30 cm diameter. In August 1970 potted plants were re-irradiated with 600 R of gamma rays (50 R/d), and the latest somatic mutation detection was made in April 1972. Somatic mutability from Kuma-sugi mutants after re-irradiation is shown in Tables 7 and 8. Mutants from re-irradiated clones were classified into two groups, waxless and other morphological mutant strains.

Table 5. Observation of pollen mother cell division in mutant strains

Kuma-sugi (Original)

IRB No.		Stages in pollen mother cell division				
		M_I	A_I	M_{II}	A_{II}	Tetrad
Kuma-sugi (Original)	Cell observed	62	26	38	45	48
	11_{II}	88.71	Nor[1] 88.46	11:11 94.74	11:11 / 11:11 93.33	Nor 93.75
	$10_{II}+1_{III}$[2]	1.61	Lag I[3] 11.53	10:12 5.26	10:10 / 10:10 2.38	Micro I[4] 6.25
	$10_{II}+L_{II}$[5]	8.06			10:2:10 / 11:11 4.44	
	$10_{II}+2_I$	1.61				

601-06

IRB No.		M_I	A_I	M_{II}	A_{II}	Tetrad
601-06	Cell observed	45	61	52	22	56
	11_{II}	46.67	Nor 77.05	11:11 78.85	11:11 / 11:11 72.73	Nor 85.71
	$10_{II}+1_{III}$	8.89	Lag I 19.67	10:12 5.77	10:2:10 / 11:11 9.09	Micro I 14.29
	$10_{II}+L_{II}$	8.89	Brid 3.28	10:1:11 11.54	9:4: 9 / 11:11 4.55	
	$10_{II}+2_I$	31.11		9:2:11 1.92	Brid 13.64	
	$9_{II}+1_{IV}$	4.44		9:4: 9 1.92		

601-22 Cell observed	75	71	48	37	58
11_{II}	54.67	Nor 90.14	11:11 77.08	11:11 / 11:11 83.78	Nor 89.66
$10_{II}+1_{IIIi}$	4.00	Lag I [6] 8.45	10:12 2.08	10:2:10 / 11:11 8.11	Micro I 10.34
$9_{II}+2_{IIIi}$	8.00	Brid 1.41	10:1:11 12.50	Frag [7] 5.41	
$10_{II}+L_{II}$	14.67		10:2:10 6.25	Brid 2.70	
$10_{II}+2_{I}$	6.67		Frag 2.08		
$9_{II}+1_{IV}$	10.67				
$11_{II}+Frag$	1.33				

Remarks : 1) Nor : Normal cell.   2) 1_{IIi} : Cell with partially homologous chromosomes.

3) Lag I : Cell with one lagging chromosome. 4) Micro I : Cell with one micro-nucleus.

5) L_{II} : Unsynchronized bivalent.   6) Brid : Cell with chromosome bridge.

7) Frag : Cell with fragment.

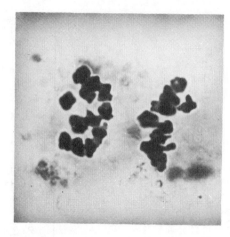

FIG.5. Unequal chromosome distribution at MII, 10:12 (IRB601-9).

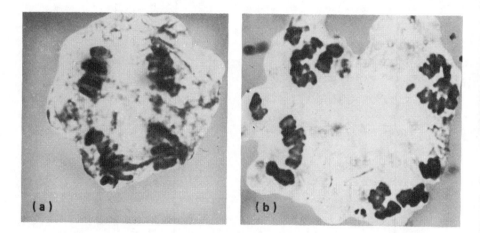

FIG.6. Chromosome abnormalities: (a) Chromosome bridge at AII (IRB601-2);
(b) Chromosome fragment at AII (IRB601-33).

3.2. Results and discussion

Concerning the somatic mutation rate, control plants of both groups showed nearly the same values, whereas re-irradiated waxless mutant strains showed a very high rate of somatic mutation, especially in IRB601-21 and 601-26. Morphological mutant strains seemed to be less able to undergo mutation in comparison with the control plants. As seen in Table 8, the type of somatic mutation was also quite different between the two groups of mutant strains.

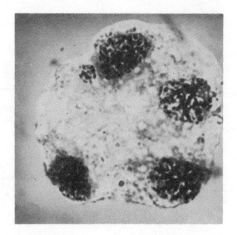

FIG.7. Micronuclei at the tetrad stage (IRB601-23).

In morphological mutant strains the lowest somatic mutation frequency was observed. On the other hand, in waxless mutant strains, morphological somatic mutations, such as fat or stout stems and thick and short needles, were induced very frequently by re-irradiation. Kuma-sugi produced chlorophyll, waxless and other morphological mutants through gamma irradiation. These mutations were also induced by re-irradiation, although other kinds of mutants were not detected. It was not apparent that there were differences of mutation frequency between waxless mutants and other morphological mutants.

4. ROOTING ABILITY IN MUTANT STRAINS OF SUGI *(Cryptomeria japonica* D. Don)

About 90 mutants of Sugi have already been detected and propagated vegetatively at the Institute of Radiation Breeding. These mutants, induced either by acute or chronic gamma irradiation, originated from the local varieties Kuma-sugi, Iwao-sugi, Aya-sugi, Sanbu-sugi, Yabukuguri-sugi, Okinoyama-sugi, Masuyama-sugi, Boka-sugi, Yame-hon-sugi, Measa-sugi, Hitoyoshi-measa-sugi and Tateyama-sugi. The rooting ability of the mutant strains was investigated.

4.1. Materials and methods

Cutting tests were made in 1975 and 1977. Cuttings were planted in sand beds in a mist-spraying greenhouse in the spring of 1975 and also in vermiculite

Table 6.   Abnormal cell frequencies in pollen mother cell division in Sugi mutant strains

IRB No.	Abnormal cell frequencies in pollen mother cell division (%)					
	M_I	A_I	M_{II}	A_{II}	Tetrad	
Kuma-sugi (Original)	11.29	11.54	5.26	6.67	6.25	
601-01	41.38	50.00	23.08	21.21	21.31	
601-02	60.94	9.52	21.54	30.30	12.82	
601-03	56.14	22.22	22.86	31.07	14.85	
601-06	53.33	22.95	21.15	27.27	14.29	
601-09	59.18	13.51	34.48	28.57	12.50	
601-17m	41.18	29.17	31.82	36.84	33.33	
601-17	43.55	6.25	8.33	7.14	8.33	
601-21	50.00	17.14	2.00	7.69	3.57	
601-22	45.33	9.86	22.92	16.22	10.34	
601-23	41.18	13.04	18.00	18.18	13.64	
601-24	42.37	0.	8.33	5.56	3.23	
601-33	56.00	24.53	31.11	36.36	23.53	
WPL 147	49.21	13.64	25.93	20.00	18.33	

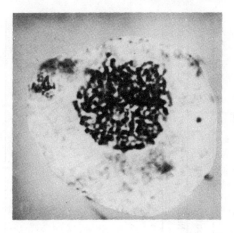

FIG.8.   Micronuclei at the pachytene stage (IRB601-9).

beds of the growth chamber in the autumn of 1977. The materials used and their
characteristics are listed in Table 1. In the cutting test in the spring of 1975
cuttings were collected from nine mutant strains and their original plant (Kuma-
sugi) and from branchlets that were not mutated morphologically after mutagenic
treatment. Cuttings were removed from the plants in the middle of April and
the cut-ends of the branchlets were dipped in running water for about 24 h.
The length of the cuttings was 15 cm and they were planted on 25 April 1975.
Irrigation above the cuttings was controlled by artificial mist spraying. The
number of rooted cuttings, rooted positions and the fresh weight of roots were
examined on 20 and 23 October 1975. In the cutting test in the autumn of 1977
cuttings were collected from 16 mutant strains, control plants from the unmutated
branchlets corresponding to each mutant and several local varieties of Sugi.
Irradiated branchlets that did not mutate morphologically and non-irradiated
original mother plants were planted as controls. Cuttings were removed on
3 November 1977, pre-treated by dipping the cut-end in running water for about
24 h and on 4 November cutting was carried out in the growth chamber. The
cutting length was 8 cm and the experiment was carried out on three replications;
each plot consisted of 16 cuttings. Cuttings were planted in plant beds
(60 × 40 × 12 cm) filled with vermiculite and covered with a sheet of vinyl film
to control evaporation. The plant beds were set under a light of about 8000 lx.
Room conditions were 16 h/d at 26°C and 22°C in the dark period. The beds
were irrigated every 3 days. The number of rooted cuttings, rooted positions
and the fresh weight of roots were examined on 13 April 1978.

Table 7. Somatic mutability of re-irradiated Kuma-sugi mutants propagated by cutting

IRB No.	Mutant				Control plant*			
	Number of plant	Number of branch	Number of mutation	Mutation rate(%)	Number of plant	Number of branch	Number of mutation	Mutation rate(%)
Waxless mutants								
601-17	3	624	10	1.60	2	424	4	0.94
601-21	2	729	48	6.58	2	490	2	0.41
601-23	4	903	28	3.10	4	1286	11	0.86
601-26	2	251	19	7.54	-	-	-	-
Morphological mutants								
601- 1	2	1042	1	0.10	1	435	5	1.15
601-15	5	715	3	0.42	4	1149	1	0.09
601- 3	8	894	2	0.22	3	1019	5	0.49
601- 6	1	122	0	0.00	3	954	5	0.52

* Control plant : Irradiated plant without any morphological changes from each mother plant of the mutant.

Irradiated branchlets that did not mutate morphologically and non-irradiated mother plants of the local varieties Kuma-sugi, Boka-sugi and Iwao-sugi (all used as controls) were irradiated with the same dose as each mutant.

4.2. Results and discussion

In the test in the spring of 1975 control plants showed a higher rooting ability than the mutant strains and Kuma-sugi (mother plant of mutants and controls). Only one mutant strain (IRB601-3) showed a lower rooting ability (22.2%) than its mother plant (24.5%). The highest rooting ability among the mutant strains was found in IRB601-2 (65.7%). In a number of roots and fresh weights of roots, no differences were observed between the mutant strains and the control, but there was an obvious difference between mother plant and mutant strains. Kuma-sugi showed a high survival rate, a low rooting ability and a low fresh weight of roots. It was thought that rooting of Kuma-sugi needed a much longer period than the mutant strains and the controls. The number of roots in IRB601-80 was very large (13.8), but the fresh weight of roots was not drastically greater because of fine root.

In the rooting ability test in the autumn of 1977 there was a distinct difference between the mutant strains and the controls (Table 9). In IRB601-2, 5, 14, 17, 18 and 65 higher rooting abilities were found than in the mother plants. Among the local varieties Iwao-sugi showed the highest rooting ability. Low rooting ability was observed in the waxless mutant strains (IRB601-21, 601-22, 601-24, 601-33 and 601-35) and in the local varieties of Kuma-sugi, Boka-sugi and Sanbu-sugi. But the longer duration of the cutting experiments on Sanbu-sugi and Boka-sugi might have influenced the greater rooting ability, since they had a high survival rate.

The rooting ability of Sugi was influenced by the unequal planting site of the mother tree, the location of branchlets and the physiological characteristics of the mother tree. From the results of the 1975 and 1977 experiments the control showed a lower rate of rooted cuttings in 1977 than in 1975, especially in the 1977 control, IRB601-14, which had the lowest rate. This shows that 8 000 lx light intensity was not sufficient for rooting and growth. These control plants, propagated vegetatively, were planted next to each mutant strain in the mutant bank of the Institute of Radiation Breeding. Therefore, the higher rooting ability of these controls is not considered to be due to any of the effective mutational events on rooting but to selection for rooting ability. In autumn cutting was performed in the dormant stage of Sugi, whereas in spring cutting branchlets were collected when the buds began to sprout, so that the former needed longer for rooting than the latter. It was thought that in IRB601-14, which had the lowest rooting ability even though it had high survival rate, the differences between natural and artificial light conditions and between the sprouting and dormant stages when the cuttings were removed influenced the rooting ability. IRB601-65 showed a

Table 8. Kind of somatic mutation induced from Kuma-sugi mutants after re-irradiation

IRB No.	Kind of somatic mutation (Mutant)				Kind of somatic mutation (Control*)			
	Chlorophyll	Morphology	Waxless	Total	Chlorophyll	Morphology	Waxless	Total
Waxless mutants								
601-17	1	9	0	10	3	1	0	4
601-21	1	47	0	48	0	2	0	2
601-23	13	15	0	28	9	1	1	11
601-26	0	19	0	19	-	-	-	-
Total	15	90	0	105	12	4	1	17
Morphological mutants								
601- 1	1	0	0	1	4	0	1	5
601-15	2	0	2	4	0	1	0	1
601- 3	0	1	1	2	4	0	1	5
601- 6	0	0	0	0	2	2	1	5
Total	3	1	3	7	10	3	3	16

* Control : Irradiated plant without any morphological changes from each mother plant of the mutant

Table 9.   Rooting ability in Sugi mutant strains (1977)

IRB No.	Control				Mutant			
	Survival rate (%)	Rate of rooted cutting (%)	Number of root	Fresh weight of root (g)	Survival rate (%)	Rate of rooted cutting (%)	Number of root	Fresh weight of root (g)
601-01	97.9	78.6	2.2	0.45	88.3	51.4	1.7	0.34
601-02	84.6	65.1	2.4	0.62	100.0	86.5	2.9	0.66
601-03	88.9	71.7	2.6	0.50	84.4	44.5	4.0	0.44
601-05	62.8	44.8	2.8	0.47	84.6	69.3	3.4	0.53
601-11	84.4	55.5	2.3	0.59	64.6	39.6	2.0	0.51
601-14	97.6	12.6	1.2	0.33	79.8	55.1	2.9	0.23
601-16	84.4	33.3	1.9	0.64	89.6	50.0	2.2	0.93
601-17	91.1	62.2	1.7	0.69	93.2	69.5	2.3	0.29
601-18	40.7	28.7	2.2	0.34	77.1	52.8	2.5	0.35
601-21	95.6	72.8	3.0	0.77	77.5	43.8	2.1	0.36
601-22	95.6	72.8	3.0	0.77	93.5	55.2	3.2	0.48
601-24	66.7	60.0	2.4	0.44	58.4	56.2	2.5	0.46
601-33	100.0	77.8	2.7	0.31	77.8	33.3	1.6	0.48
601-35	74.4	61.4	3.1	0.53	51.5	40.8	2.9	0.51
601-80	73.3	51.8	2.1	0.44	74.0	46.8	2.8	0.50
Kuma-sugi	62.5	43.8	1.8	0.38				
601-65 (Boka-sugi)	95.4	36.4	2.3	0.21	100.0	80.6	3.2	0.51
Sanbu-sugi	97.8	28.7	2.8	0.31				
Iwao-sugi	100.0	85.3	2.8	0.27				

high rooting ability (80.6%) and all the cuttings of this mutant strain, which had formed calluses by the end of this experiment (19.4%), rooted after 21 d. It was considered that the defective rooting ability in waxless mutant strains was mainly due to the unbalanced absorption and evaporation of water. Since rooting ability is influenced by various environmental conditions of the cutting and the mother plant and also by its growth habit, sufficiently replicated experiments are needed with different mother plants, in order to vary their growing conditions and planting sites.

REFERENCES

[1] OHBA, K., Studies on radiosensitivity and induction of somatic mutation in forest trees, Gamma Field Symp. 3 (1964) 111.

[2] OHBA, K., Studies on the radiation breeding of forest trees, Bull. Inst. Radiat. Breed. 2 (1971) 1.

[3] OHBA, K., MURAI, M., Varietal difference in somatic mutability of Sugi, *Cryptomeria japonica* D. Don, Gamma Field Symp. 8 (1969) 109.

[4] OHBA, K., MURAI, M., Effect of gamma ray irradiation, pruning and internal disbudding on induction of somatic mutation in Kuma-sugi, *Cryptomeria japonica* D. Don, J. Jpn. Forestry 53 (1971) 170 (in Japanese).

[5] ALISTON, R.E., SPARROW, A.H., Somatic mutation rates in double and triple heterozygotes of *Impatiens balsamina* following chronic gamma irradiation, Radiat. Bot. 1 (1962) 229.

[6] NYBOM, N., KOH, A., Induced mutations and breeding methods in vegetatively propagated plants, Suppl. Radiat. Bot. 5 (1965) 661.

[7] OHBA, K., MAETA, T., Induction of somatic mutation and the mutants in Sugi, *Cryptomeria japonica* D. Don, Gamma Field Symp. 12 (1973) 19.

[8] NISHIDA, T., NAKAJIMA, K., TAKATO, S., Radiosensitivity and induction of somatic mutations in woody perennials under chronic gamma irradiation, Gamma Field Symp. 6 (1967) 19.

FINAL REVIEW OF THE
CO-ORDINATED RESEARCH PROGRAMME

1. BACKGROUND AND SCOPE

Many important plant species for food and for industrial use are vegetatively
propagated and woody perennials with long generation cycles. The lack or
difficulty of sexual propagation in these plant species limit or exclude the
possibility of creating new plant types through crosses. Mutation techniques
by using ionizing radiations and other mutagens provide a good opportunity to
produce new variability in such species from which new plant types can be
selected with genetically improved characteristics and properties.

A panel of experts convened by the Joint FAO/IAEA Division of Atomic
Energy in Food and Agriculture in Vienna in 1972 concluded that successful use
of induced mutations had been obtained with ornamental plants and certain
fruit trees (apples, cherries), and that similar results might be achieved in other
vegetatively propagated plant species such as citrus, mulberry, palm, sugar-cane,
banana, forage grasses, tuber crops, etc.

On the basis of this advice a co-ordinated research programme was established by the IAEA in 1972, involving 14 research institutes in developing countries
and five in developed countries. Biennial research co-ordination meetings were
held (Japan 1974, Netherlands 1976, Poland 1978, India 1980).

The research aims of the programme were to develop techniques for induction,
selection and utilization of mutants in various vegetatively propagated crops and
tree crops as recommended by the panel of experts in September 1972.

2. PROGRESS AND ACHIEVEMENTS

The programme has included a wide variety of plant species of cultivated
plants and trees in tropical and temperate climates, namely: sugar-cane, sweet
potato, potato, cassava, apple, cherry, grape, olive, peach, mulberry, ornamental
plants and forage grasses. With regard to the technique of mutation induction
and use of mutagenic agents, consistent efforts have been made and good methods
are now available especially in ornamental plants and in temperate zone fruit
crops. More extensive studies in other crops and especially in relation to the
rapid development of in vitro techniques are advisable. For mutant selection it is
necessary to have a suitable methodology for each type of crop, according to its
method of propagation. Such methods are now established in many of the
investigated crops. In temperate zone fruit crops several mutants with improved
characteristics have been selected. For example, in sweet cherry seven compact

mutants have been induced in outstanding varieties. They are of particular interest for practical use because of their easy harvesting and early entry into production. They are already agronomically evaluated and will soon be released for commercial use. In apple varieties and rootstocks, in sour cherry and in grape, several interesting mutants concerning plant size, 'compact type', fruit size, fruit skin colour and fruit taste have been isolated. Confirming trials and analyses are necessary to make definitive conclusions.

An early ripening type and a nectarine-type mutant have been induced in a peach variety. They will be released under the names 'Sprinta' and 'Rubella' respectively.

In tropical fruit radiation experiments with citrus, coconut and bananas have been performed. Tissue culture techniques have been advanced. In citrus, screening methods for saline tolerance and for resistance to 2, 4-D (herbicide) at an early stage of development have been established. In vitro plantlet formation has been achieved in coconut and banana. In sugar-cane, several mutants of economic importance have been released. They are characterized by high yield, better sugar content and/or smut resistance. One mutant which has a shorter growth period than the original variety but higher yield is now under agronomical evaluation. A non-flowering mutant with higher yields than the parental variety has also good prospects.

Among tuber crops, a suitable in vitro propagation technique has been established for potato starting from leaf explants. From irradiated leaves a large number of adventitious plantlets can be raised which are in general solid, non-chimeric mutants. Attempts to regenerate plants from single cells isolated from callus and leaf mesophyll were also undertaken. Screening for mutations resistant to cyst nematodes lead to the isolation of one mutant which is now under further investigation. Studies on sweet potatoes were begun with the aim of improving by mutagenesis qualitative and quantitative characters such as tuber production, dry matter content and sugar content. Some mutants were detected with improved characteristics in comparison with the original varieties.

In mulberry, a suitable method of conidia inoculation has been developed for selecting mutants resistant to *Diaporthe nomurai* H. ('die-back disease'). So far, however, no mutants have been found. Rhizome gamma irradiation of sterile hybrids in forage grasses (Bermuda grass) has proved to be a suitable method for inducing mutants. Among the mutants with improved characteristics are those with better adaptability, nematode resistance or tolerance, and increased winter hardiness. Depending on their performance in large-scale field trials which are now being carried out, the mutants may soon be released for commercial use.

Mutagenesis applied in ornamental plants has been very successful. By using the in vitro propagation technique through adventitious bud formation it has been possible to obtain almost solid (non-chimeric) mutants. There is certainly a great potential for the application of in vitro techniques in other vegetatively propagated plants. Many mutants induced in chrysanthemum, begonia, dianthus, achimenes, streptocarpus and other species have been released for commercial use.

3.   CONCLUSIONS AND RECOMMENDATIONS

The aims of this co-ordinated programme have been to develop mutation breeding techniques and to produce desirable variability from which new plant types with improved properties can be selected. The lack or difficulty of sexual propagation in the vegetatively propagated species excludes or limits the possibility of creating new and improved plant genotypes through cross breeding. Consequently, in the past plant improvement has depended largely upon the selection of naturally occurring mutants. Irradiation techniques that increase the frequency of mutations can be regarded as crucial for accelerating the genetic improvement of this group of plants.

The programme has involved qualified research institutes in developing and developed countries. Results of great economical importance have been achieved in those plants where a suitable treatment technique as well as proper methodology of somatic mutation isolation have been available. Judging from the encouraging results obtained within the frame of this programme, it appears justified to consider the establishment of a follow-up programme focusing specifically on tropical crop plants of importance for developing countries. These crops may include coconut, cardamom, cassava, sweet potato, potato, sugar-cane, citrus, banana, pineapple and olive. The work to be undertaken would benefit from co-operation between plant geneticists, plant breeders, plant physiologists, phytopathologists and in vitro culture specialists.

The research aims should be concerned with developing efficient methods for the induction, selection and evaluation of mutants. Specifically, such a programme should include the following problems:

Choice of the mutagenic agent, and optimal treatment methodology, both in relation to the further development of in vitro techniques
More effective screening methods for early detection and isolation of desirable mutants
Better knowledge of the possibility to select at the cellular level desirable mutant types that can be expressed in the whole plant
Extension of the use of cell culture for induction and recovery of solid mutants to other crops.

LIST OF PARTICIPANTS

Broertjes, C.

Institute for Atomic Sciences in Agriculture,
6 Keyenbergseweg, Postbus 48, Wageningen,
The Netherlands

Burton, G.W.

USDA, Federal Research,
Southern Region, Coastal Plain Station,
Tifton, GA 31794, United States of America

Donini, B.

Laboratorio Applicazioni in Agricoltura,
Comitato Nazionale per l'Energia Nucleare,
Centro di Studi Nucleari della Casaccia,
S. Maria di Galeria, I-00060 Rome, Italy

Gröber, K.

Zentralinstitut für Genetik und Kulturpflanzenforschung,
Akademie der Wissenschaften der DDR,
Corrensstrasse 33, DDR-4325 Gatersleben,
German Democratic Republic

Grodzinsky, D.M.

Institute of Plant Physiology of the
    Ukrainian SSR Academy of Science,
Kiev 32, USSR

de Guzman, V.

College of Agriculture,
University of the Philippines at Los Banos,
Laguna, Philippines

Jagathesan, D.

Sugar-Cane Breeding Institute,
Coimbatore 641007 (Tamil Nadu),
India

Kukimura, H.

Institute of Radiation Breeding,
P.O. Box 3, Ohmiya-machi,
Naka-gun, Ibaraki-ken, Japan

Lacey, C.N.D.

Long Ashton Research Station,
Long Ashton, Bristol, Avon BS18 PAF,
United Kingdom

Nakajima, K.

Sericultural Experiment Station, M.A.F.F.,
55–30, Wada 3-chome, Suginami-ku,
Tokyo 166, Japan

Privalov, G.F.	Academy of Sciences of the USSR, Siberian Department, Institute of Cytology and Genetics, 90 prospekt Nauki, 10, 630090 Novosibirsk, USSR
Siddiqui, S.H.	Atomic Energy Agricultural Research Centre, Tandojam, Pakistan
Upadhya, M.D.	Central Potato Research Institute, Simla 1 (Himachal Pradesh), India
Zagaja, S.W.	Research Institute of Pomology and Floriculture, 96–100 Skierniewice, Pomologiczna 18, Poland
Zubrzycki, H.M.	Mejoramiento Cítrico, Instituto Nacional de Tecnología Agropecuaria, Estación Experimental Agropecuaria, Bella Vista (Corrientes), Argentina
Mikaelsen, K. *(Scientific Secretary)*	Joint FAO/IAEA Division of Isotope and Radiation Applications of Atomic Energy for Food and Agricultural Development, Wagramerstrasse 5, P.O. Box 100, A-1400 Vienna, Austria

The following conversion table is provided for the convenience of readers

FACTORS FOR CONVERTING SOME OF THE MORE COMMON UNITS TO INTERNATIONAL SYSTEM OF UNITS (SI) EQUIVALENTS

NOTES:
(1) SI base units are the metre (m), kilogram (kg), second (s), ampere (A), kelvin (K), candela (cd) and mole (mol).
(2) ▶ indicates SI derived units and those accepted for use with SI;
   ▷ indicates additional units accepted for use with SI for a limited time.
   [*For further information see the current edition of The International System of Units (SI), published in English by HMSO, London, and National Bureau of Standards, Washington, DC, and International Standards ISO-1000 and the several parts of ISO-31, published by ISO, Geneva.*]
(3) The correct symbol for the unit in column 1 is given in column 2.
(4) ✻ indicates conversion factors given exactly; other factors are given rounded, mostly to 4 significant figures:
   ≡ indicates a definition of an SI derived unit: [ ] in columns 3+4 enclose factors given for the sake of completeness.

Column 1 *Multiply data given in:*	Column 2	Column 3 *by:*	Column 4 *to obtain data in:*	
Radiation units				
▶ becquerel	1 Bq	(has dimensions of s^{-1})		
disintegrations per second (= dis/s)	$1\ s^{-1}$	$\equiv 1.00 \times 10^0$	Bq	✻
▷ curie	1 Ci	$= 3.70 \times 10^{10}$	Bq	✻
▷ roentgen	1 R	$[= 2.58 \times 10^{-4}$	C/kg]	✻
▶ gray	1 Gy	$[\equiv 1.00 \times 10^0$	J/kg]	✻
▷ rad	1 rad	$= 1.00 \times 10^{-2}$	Gy	✻
▶ sievert *(radiation protection only)*	1 Sv	$[= 1.00 \times 10^0$	J/kg]	✻
rem *(radiation protection only)*	1 rem	$[= 1.00 \times 10^{-2}$	J/kg]	✻
Mass				
▶ unified atomic mass unit ($\frac{1}{12}$ of the mass of ^{12}C)	1 u	$[= 1.660\ 57 \times 10^{-27}$	kg, approx.]	
▶ tonne (= metric ton)	1 t	$[= 1.00 \times 10^3$	kg]	✻
pound mass (avoirdupois)	1 lbm	$= 4.536 \times 10^{-1}$	kg	
ounce mass (avoirdupois)	1 ozm	$= 2.835 \times 10^1$	g	
ton (long) (= 2240 lbm)	1 ton	$= 1.016 \times 10^3$	kg	
ton (short) (= 2000 lbm)	1 short ton	$= 9.072 \times 10^2$	kg	
Length				
statute mile	1 mile	$= 1.609 \times 10^0$	km	
nautical mile (international)	1 n mile	$= 1.852 \times 10^0$	km	✻
yard	1 yd	$= 9.144 \times 10^{-1}$	m	✻
foot	1 ft	$= 3.048 \times 10^{-1}$	m	✻
inch	1 in	$= 2.54 \times 10^1$	mm	✻
mil (= 10^{-3} in)	1 mil	$= 2.54 \times 10^{-2}$	mm	✻
Area				
▷ hectare	1 ha	$[= 1.00 \times 10^4$	m^2]	✻
▷ barn *(effective cross-section, nuclear physics)*	1 b	$[= 1.00 \times 10^{-28}$	m^2]	✻
square mile, (statute mile)^2	1 mile^2	$= 2.590 \times 10^0$	km^2	
acre	1 acre	$= 4.047 \times 10^3$	m^2	
square yard	1 yd^2	$= 8.361 \times 10^{-1}$	m^2	
square foot	1 ft^2	$= 9.290 \times 10^{-2}$	m^2	
square inch	1 in^2	$= 6.452 \times 10^2$	mm^2	
Volume				
▶ litre	1 l *or* 1 ltr	$[= 1.00 \times 10^{-3}$	m^3]	✻
cubic yard	1 yd^3	$= 7.646 \times 10^{-1}$	m^3	
cubic foot	1 ft^3	$= 2.832 \times 10^{-2}$	m^3	
cubic inch	1 in^3	$= 1.639 \times 10^4$	mm^3	
gallon (imperial)	1 gal (UK)	$= 4.546 \times 10^{-3}$	m^3	
gallon (US liquid)	1 gal (US)	$= 3.785 \times 10^{-3}$	m^3	

This table has been prepared by E.R.A. Beck for use by the Division of Publications of the IAEA. While every effort has been made to ensure accuracy, the Agency cannot be held responsible for errors arising from the use of this table.

Column 1 *Multiply data given in:*		Column 2	Column 3 *by:*	Column 4 *to obtain data in:*

Velocity, acceleration

foot per second (= fps)	1 ft/s	= 3.048 × 10^{-1}	m/s	*
foot per minute	1 ft/min	= 5.08 × 10^{-3}	m/s	*
mile per hour (= mph)	1 mile/h	= $\begin{cases} 4.470 \times 10^{-1} & \text{m/s} \\ 1.609 \times 10^{0} & \text{km/h} \end{cases}$		
▷ knot (international)	1 knot	= 1.852 × 10^{0}	km/h	*
free fall, standard, g		= 9.807 × 10^{0}	m/s^2	
foot per second squared	1 ft/s^2	= 3.048 × 10^{-1}	m/s^2	*

Density, volumetric rate

pound mass per cubic inch	1 lbm/in^3	= 2.768 × 10^4	kg/m^3
pound mass per cubic foot	1 lbm/ft^3	= 1.602 × 10^1	kg/m^3
cubic feet per second	1 ft^3/s	= 2.832 × 10^{-2}	m^3/s
cubic feet per minute	1 ft^3/min	= 4.719 × 10^{-4}	m^3/s

Force

▶ newton	1 N	[≡ 1.00 × 10^0	m·kg·s^{-2}]	*
dyne	1 dyn	= 1.00 × 10^{-5}	N	*
kilogram force (= kilopond (kp))	1 kgf	= 9.807 × 10^0	N	
poundal	1 pdl	= 1.383 × 10^{-1}	N	
pound force (avoirdupois)	1 lbf	= 4.448 × 10^0	N	
ounce force (avoirdupois)	1 ozf	= 2.780 × 10^{-1}	N	

Pressure, stress

▶ pascal	1 Pa	[≡ 1.00 × 10^0	N/m^2]	*
▷ atmosphere ^{a}, standard	1 atm	= 1.013 25 × 10^5	Pa	*
▷ bar	1 bar	= 1.00 × 10^5	Pa	*
centimetres of mercury (0°C)	1 cmHg	= 1.333 × 10^3	Pa	
dyne per square centimetre	1 dyn/cm^2	= 1.00 × 10^{-1}	Pa	*
feet of water (4°C)	1 ftH$_2$O	= 2.989 × 10^3	Pa	
inches of mercury (0°C)	1 inHg	= 3.386 × 10^3	Pa	
inches of water (4°C)	1 inH$_2$O	= 2.491 × 10^2	Pa	
kilogram force per square centimetre	1 kgf/cm^2	= 9.807 × 10^4	Pa	
pound force per square foot	1 lbf/ft^2	= 4.788 × 10^1	Pa	
pound force per square inch (= psi) ^{b}	1 lbf/in^2	= 6.895 × 10^3	Pa	
torr (0°C) (= mmHg)	1 torr	= 1.333 × 10^2	Pa	

Energy, work, quantity of heat

▶ joule (≡ W·s)	1 J	[≡ 1.00 × 10^0	N·m]	*
▶ electronvolt	1 eV	[= 1.602 19 × 10^{-19}	J, approx.]	
British thermal unit (International Table)	1 Btu	= 1.055 × 10^3	J	
calorie (thermochemical)	1 cal	= 4.184 × 10^0	J	*
calorie (International Table)	1 cal$_{IT}$	= 4.187 × 10^0	J	
erg	1 erg	= 1.00 × 10^{-7}	J	*
foot-pound force	1 ft·lbf	= 1.356 × 10^0	J	
kilowatt-hour	1 kW·h	= 3.60 × 10^6	J	*
kiloton explosive yield (PNE) (≡ 10^{12} g-cal)	1 kt yield	≃ 4.2 × 10^{12}	J	

---

^{a} atm (g)　(= atü): atmospheres gauge
　　atm abs　(= ata): atmospheres absolute

^{b} lbf/in^2 (g)　(= psig): gauge pressure;
　　lbf/in^2 abs　(= psia): absolute pressure.

Column 1 Multiply data given in:	Column 2	Column 3 by:	Column 4 to obtain data in:

Power, radiant flux

▶ watt	1 W	$[\equiv 1.00 \times 10^0$	J/s]	*
British thermal unit (International Table) per second	1 Btu/s	$= 1.055 \times 10^3$	W	
calorie (International Table) per second	1 cal$_{IT}$/s	$= 4.187 \times 10^0$	W	
foot-pound force/second	1 ft·lbf/s	$= 1.356 \times 10^0$	W	
horsepower (electric)	1 hp	$= 7.46 \times 10^2$	W	*
horsepower (metric) (= ps)	1 ps	$= 7.355 \times 10^2$	W	
horsepower (550 ft·lbf/s)	1 hp	$= 7.457 \times 10^2$	W	

Temperature

▶ kelvin — K — — — — — — — — — — — — — — — — — —

▶ degrees Celsius, t $t = T - T_0$ *

    where T is the thermodynamic temperature in kelvin
    and T_0 is defined as 273.15 K

degree Fahrenheit	$t_{°F} - 32$
degree Rankine	$T_{°R}$
temperature difference[c]	$\Delta T_{°R} (= \Delta t_{°F})$

$\left. \right\} \times \left(\dfrac{5}{9}\right)$ gives $\left\{ \begin{array}{l} t \text{ (in degrees Celsius)} \ *\\ T \text{ (in kelvin)} \ *\\ \Delta T (= \Delta t) \ * \end{array} \right.$

— — — — — — — — — — — — — — — — — — —

Thermal conductivity [c]

1 Btu·in/(ft^2·s·°F)	(International Table Btu)	$= 5.192 \times 10^2$	W·m^{-1}·K^{-1}
1 Btu/(ft·s·°F)	(International Table Btu)	$= 6.231 \times 10^3$	W·m^{-1}·K^{-1}
1 cal$_{IT}$/(cm·s·°C)		$= 4.187 \times 10^2$	W·m^{-1}·K^{-1}

Miscellaneous quantities

litre per mole per centimetre	(1M/cm =) 1 ltr·mol^{-1}·cm^{-1}	$= 1.00 \times 10^{-1}$ m^2/mol	*	
(molar extinction coefficient or molar absorption coefficient)				
G-value, traditionally quoted per 100 eV				
of energy absorbed	1×10^{-2} eV^{-1}	$= 6.24 \times 10^{16}$	J^{-1}	
(radiation yield of a chemical substance)				
mass per unit area	1 g/cm^2	$[= 1.00 \times 10^1$	kg/m^2]	*
(absorber thickness and mean mass range)				

---

[c] A temperature interval or a Celsius temperature difference can be expressed in degrees Celsius as well as in kelvins.

HOW TO ORDER IAEA PUBLICATIONS

An exclusive sales agent for IAEA publications, to whom all orders and inquiries should be addressed, has been appointed in the following country:

UNITED STATES OF AMERICA   UNIPUB, 345 Park Avenue South, New York, NY 10010

In the following countries IAEA publications may be purchased from the sales agents or booksellers listed or through your major local booksellers. Payment can be made in local currency or with UNESCO coupons.

ARGENTINA	Comisión Nacional de Energía Atomica, Avenida del Libertador 8250, RA-1429 Buenos Aires
AUSTRALIA	Hunter Publications, 58 A Gipps Street, Collingwood, Victoria 3066
BELGIUM	Service Courrier UNESCO, 202, Avenue du Roi, B-1060 Brussels
CZECHOSLOVAKIA	S.N.T.L., Spálená 51, CS-113 02 Prague 1
	Alfa, Publishers, Hurbanovo námestie 6, CS-893 31 Bratislava
FRANCE	Office International de Documentation et Librairie, 48, rue Gay-Lussac, F-75240 Paris Cedex 05
HUNGARY	Kultura, Hungarian Foreign Trading Company P.O. Box 149, H-1389 Budapest 62
INDIA	Oxford Book and Stationery Co., 17, Park Street, Calcutta-700 016
	Oxford Book and Stationery Co., Scindia House, New Delhi-110 001
ISRAEL	Heiliger and Co., Ltd., Scientific and Medical Books, 3, Nathan Strauss Street, Jerusalem 94227
ITALY	Libreria Scientifica, Dott. Lucio de Biasio "aeiou", Via Meravigli 16, I-20123 Milan
JAPAN	Maruzen Company, Ltd., P.O. Box 5050, 100-31 Tokyo International
NETHERLANDS	Martinus Nijhoff B.V., Booksellers, Lange Voorhout 9-11, P.O. Box 269, NL-2501 The Hague
PAKISTAN	Mirza Book Agency, 65, Shahrah Quaid-e-Azam, P.O. Box 729, Lahore 3
POLAND	Ars Polona-Ruch, Centrala Handlu Zagranicznego, Krakowskie Przedmiescie 7, PL-00-068 Warsaw
ROMANIA	Ilexim, P.O. Box 136-137, Bucarest
SOUTH AFRICA	Van Schaik's Bookstore (Pty) Ltd., Libri Building, Church Street, P.O. Box 724, Pretoria 0001
SPAIN	Diaz de Santos, Lagasca 95, Madrid-6
	Diaz de Santos, Balmes 417, Barcelona-6
SWEDEN	AB C.E. Fritzes Kungl. Hovbokhandel, Fredsgatan 2, P.O. Box 16356, S-103 27 Stockholm
UNITED KINGDOM	Her Majesty's Stationery Office, Agency Section PDIB, P.O. Box 569, London SE1 9NH
U.S.S.R.	Mezhdunarodnaya Kniga, Smolenskaya-Sennaya 32-34, Moscow G-200
YUGOSLAVIA	Jugoslovenska Knjiga, Terazije 27, P.O. Box 36, YU-11001 Belgrade

---

Orders from countries where sales agents have not yet been appointed and requests for information should be addressed directly to:

Division of Publications
International Atomic Energy Agency
Wagramerstrasse 5, P.O. Box 100, A-1400 Vienna, Austria